Nutritional Neurosciences

Series Editor

Mohamed Essa, Sultan Qaboos University, Qaboos, Oman

This book series aims to publish volumes focusing on both basic and clinical research in the field of nutritional neuroscience with a focus on delineating the effect of nutrition on brain function and behavior. The books will examine the role of different nutrients, food agents and supplements (both macro and micro) on brain health, neurodevelopment, neurochemistry, and behaviour. The books will examine the influence of diet, including phytochemicals, antioxidants, dietary supplements, food additives, and other nutrients on the physiology and metabolism of neurons, neurotransmitters and their receptors, cognition, behavior, and hormonal regulations.

The books will also covers the influence of nutrients and dietary supplements on the management of neurological disorders. It details the mechanism of action of phytonutrients on signaling pathways linked with protein folding, aggregation, and neuroinflammation. The books published in the series will be useful for neuroscientists, nutritionists, neurologists, psychiatrists, and those interested in preventive medicine.

Mohamed Salama
Editor

Nutrigenomics and the Brain

Springer

Editor
Mohamed Salama
The Institute of Global Public Health and
Human Ecology
American University in Cairo
New Cairo, Egypt

ISSN 2730-6712 ISSN 2730-6720 (electronic)
Nutritional Neurosciences
ISBN 978-981-16-9204-8 ISBN 978-981-16-9205-5 (eBook)
https://doi.org/10.1007/978-981-16-9205-5

This Springer imprint is published by the registered company Springer Nature Singapore Pte Ltd.
The registered company address is: 152 Beach Road, #21-01/04 Gateway East, Singapore 189721, Singapore

Contents

About the Editor

Mohamed Salama is Atlantic Senior Fellow for Equity in Brain Health at the Global Brain Health Institute (GBHI) and Associate professor at the Institute of Global Health and Human Ecology at the American University in Cairo (AUC). He established the first Translational Neuroscience Unit in Egypt. Mohamed's collaborative research led to establishing the Egyptian Network for Neurodegenerative Disorders. Mohamed was selected as a SOT Global Senior Scholar in 2013 and Translational/bridging awardee in 2016. He was awarded by Parkinson's and Movement Disorders Foundation (PMDF) for his continued research in neurodegeneration. Recently, Mohamed and his colleagues succeeded in drafting the first Reference Egyptian Genome and collaborating with other colleagues to start a national cohort (A Longitudinal Study of Egyptian Healthy Aging [AL-SEHA]).

Chapter 1
Nutrigenomics: A Broader Concept

Mohamed Salama

Nutrigenomics is one aspect of the interaction between environment and genes. The term exceeds the conventional understanding of the genes affecting body reaction to diet into more holistic approach, taking into consideration the complexity of exposure factors as well as intrinsic responses. So, although the diet type plays an important role, the external "exposure" is not limited to but extends to include lifestyle, air quality, environmental pollutants, stress, and different comodifiers that forms totally and exposome complex pattern.

On the other hand, intrinsic factor is not limited to "genetic polymorphism"; rather it extends to the complexity of several OMICS responses, e.g., proteomics, transcriptomics, metabolimics, and genomics. For that, we believe that this book on "nutrigenomics" is an advanced trial to understand the complexity of such interaction, abandoning the old fashion approaches of single straight forward interaction to adopt the new paradigm of complex multilevel interactions that shape our health/disease status.

This broad understanding of nutrigenomics can be seen reflected on the different contents contributed by authors from diverse backgrounds. The complexity of brain diseases and health issues, the multiple interactions and relations that control the development of such disorders, necessitates a multidisciplinary approach that helps provide a better understanding of the complex disease processes and identify possible targets for prevention, diagnosis, and treatment.

The first chapter by Hanaa Abdelzaher is introducing the concept of gene environment interaction and how such a dynamic crosstalk could affect brain well-being.

The second chapter by Salama and Dakhlallah is giving us an important introduction to the complexity of gene nutrition interaction, describing how nutrients can

M. Salama (✉)
Institute of Global Health and Human Ecology (I-GHHE), The American University in Cairo, New Cairo, Egypt
e-mail: mohamed-salama@aucegypt.edu

M. Salama (ed.), *Nutrigenomics and the Brain*, Nutritional Neurosciences,
https://doi.org/10.1007/978-981-16-9205-5_1

affect genes and how genes can shape our response to nutrients. As we move forward, we can see the interactions and their role in specific disease processes, e.g., AD, PD, and mental disorders.

A more specialized discussion has been presented by Russell and colleagues, focusing on the autophagy-related genes and how diet style can be manipulated. Bobo et al., on the other hand, took us in a journey to differentiate between the two major components, namely, nutrigenomics and nutrigenetics.

Digging deeper into specific conditions, the chapter by Palminha and colleagues discusses the whole process of aging and its accompanying multi-morbidities and how nutrition/gene interaction can affect that. On the other hand, some of the guest authors preferred to tackle the issue from the nutritional side like vitamin D dilemma by El-Gamal and his colleagues or food additives by El-Shafie and Kakarougkas.

I believe that this book is contributing—remarkably—to the field of gene-environment interaction and will raise a group of questions that can be answered only through more research by the scientific community.

Chapter 2
The Gene-Environment Nexus: A Holistic Approach to Neurodegenerative Diseases

Hana Abdelzaher

2.1 Introduction

Neurodegenerative diseases are one of the biggest health burdens worldwide; they affect millions of people and often have poor prognosis and decreased quality of life (Mayeux 2003). This group of diseases is characterized by the fact that they are degenerative conditions that affect areas of the central nervous system leading to cognitive or motor impairments depending on the type of nervous cells that are affected by the degeneration (Minghetti 2005). The majority of these diseases have a strong correlation with age and are thus becoming a problem especially in aging populations as their frequency increases with the increase of life expectancy (Logroscino and Tortelli 2015). The reason why this group of diseases is characterized by its heterogeneity is the fact that some of them are complex and multifactorial in nature such as Alzheimer's disease, Parkinson's disease, and amyotrophic lateral sclerosis, while others result from single-gene mutations such as Huntington disease and finally others such as prion disease that can be either hereditary, sporadic, or transmitted (Pihlstrøm et al. 2018). Research has shown that certain causative genes can account for inherited forms of ND diseases and that susceptibility genes can also play a major role in sporadic development of the diseases. Time and time again, research has shown that the environment also plays a major part in the development of such diseases as exposure to metals and pesticides, nutrition, head trauma, and certain infectious diseases that can highly increase the risk of developing indeed disorders (Brown et al. 2005). Furthermore, the interaction between environmental exposure and preexisting genetic variants has been established as one of the risk factors as well. Studies performed on Twins (Gatz and Pedersen 2013) as well as

H. Abdelzaher (✉)
Institute of Global Health and Human Ecology (I-GHHE), The American University in Cairo, Egypt, New Cairo, Egypt
e-mail: hana-abdelzaher@aucegypt.edu

M. Salama (ed.), *Nutrigenomics and the Brain*, Nutritional Neurosciences,
https://doi.org/10.1007/978-981-16-9205-5_2

animal models are also highly suggestive that the gene-environment nexus can actually strongly impact the way newer degenerative diseases present, manifest, and progress (Hannan 2004). In addition, the role of the mitochondria and oxidative damage in such diseases has been repeatedly demonstrated, and various studies have attempted to clarify whether there is an association between mitochondrial DNA variants and susceptibility to development of ND (Reddy 2009).

2.2 Genetics and Neurodegenerative Diseases

The development in the insights regarding the link between genetics and neurodegenerative disorders reflects the rapid progress of techniques in the field of molecular biology. In fact, the identification of the genes causing Huntington disease in 1983 was one of the pioneering endeavors in identifying the relationship between genetics and disease (Gusella et al. 1983). This success was based on an old technique called restriction fragment length polymorphism (RFLP) that showed that Huntington's patients had a certain marker on their chromosomes (huntingtin gene, p arm of chromosome 4). With further advancements in the field of molecular biology, scientists were able to identify that this marker was actually due to an increase in the number of trinucleotide repeats in the HTT gene. This process of identifying the exact mutation took around 10 years, showcasing the difficulty of mapping mutations back then. However, this opened up the way for scientists to later on identify many causative/associated genes and mutations for various illnesses (Bates 2005). The technique used for decades later on was called linkage analysis which aims to determine the association between a certain disease and various chromosomal markers. After this linkage, researchers would narrow down the areas of genes that were actually linked with the disease. Thanks to this technique, a multitude of disease-causing mutations was identified for diseases such as prion disease, familial Alzheimer's disease, and spinocerebellar ataxia 1. This identification of disease-causing genes allowed for screening of a large number of patients with similar phenotype for the presence of the same genes, allowing researchers to begin testing hypothesis regarding etiology, frequency of said mutations, and the degree to which they can be linked to the disease (Pihlstrøm et al. 2018). However, these early strategies, although successful for model diseases that follow mendelian rules of inheritance, were not enough to identify causative mutations in complex sporadic disorders. Not all cases, for example, followed the clear-cut influence of the apolipoprotein E variant whose presence greatly increases the risk for familial Alzheimer's disease (2–3 times). This variant was fairly common and largely influential which is not the case for most variants that are associated with other complex disorders. The study of such variants required large-scale research studies, computational power, and breakthroughs in the field which luckily enough began appearing as we progressed to the twenty-first century (Hirschhorn and Daly 2005).

2.2.1 Monogenic Forms of Neurodegenerative Diseases: Alzheimer's and Parkinson's Diseases as Examples

Alzheimer's disease: Although most AD cases are complex and sporadic in nature, there exists three familial forms of the disease that follow the rules of Mendelian inheritance. These familial forms are autosomal dominant and result from the inheritance of mutations affecting the formation of amyloid b which leads to the formation of amyloid plaques. The amyloid precursor protein (APP) is expressed and found on the transmembranes of neurons. This protein is later on cleaved and released through the action of a group of enzymes known as secretases (mostly a-secretase). When cleaved, fragments known as Ab40 and Ab42 are produced, and these fragments are highly prone to aggregation. Disease-causing point mutations tip the balance in the Ab40:Ab42 ratio, causing the aggregation of Ab42. Furthermore, duplication of the gene encoding APP has been linked to susceptibility to AD with cerebral amyloid angiopathy. Mutations in genes other than APP, like PSEN1 and PSEN2 (encoding g-secretase), have also been implicated in familial AD, with PSEN1 being the most common mutation followed by APP and then PSEN2 (Pilotto et al. 2013).

Parkinson's disease: Similarly to Alzheimer's disease, mutations in the SNCA gene encoding alpha synuclein have been shown to cause familial autosomal-dominant Parkinson's disease. This was first demonstrated in 1997, and since then a multitude of mutations in this gene have been identified and demonstrated to cause early-onset poor prognosis parkinsonism. The protein, which is the subject of these mutations, has been shown to be a component of intracellular Lewy bodies which play a major role in the development of Parkinson's disease. The identified point mutations increase the protein's capacity to oligomerize, causing the formation of fibrils. Similar to Alzheimer's disease, the multiplication of the SNCA gene also causes familial autosomal-dominant Parkinson's disease. It has been shown that duplications of the gene cause a familial phenotype that resembles the sporadic one, while the presence of the gene in triplicate form causes early onset and strong cognitive impairments, showing that the amount of the wild-type protein can cause poor prognosis (Alkanli and Ay 2019). Other genes have also been linked to the disease both in autosomal-dominant and autosomal-recessive inheritance patterns. These genes include LRRK2 and VPS35; their exact mechanism of action has not been fully elucidated; however, they have been linked to various critical pathways such as autophagy and protein degradation (Klein and Westenberger 2012).

By looking at the two monogenic forms of Alzheimer's and Parkinson's disease, scientists have been able to infer important insights. Firstly, a mutation in a single gene often produces a disease phenotype that is not distinguishable from sporadic illness. Furthermore, even the familial diseases that present with a somewhat universal clinical presentation and inheritance often are the result of different mutations in different genes that affect related pathways. To further add to the complexity, a single gene has the capacity of inducing various phenotypic presentations of a disease. The studies that were discussed previously showed that protein aggregation is a strong deterministic factor in the case of neurodegenerative diseases. The two examples that were discussed show that neurodegenerative diseases can result from structural mutations or expression changes in proteins. These monogenic studies have paved the way for later on mechanistic studies to elucidate the exact role of said genes in pathogenesis (Bekris et al. 2010).

2.3 The Environment and Its Effect

The belief that environmental exposures are strongly implicated in neurodegeneration is not a new one; environmental exposures such as certain types of food, metals, contaminants, pathogenic, and nonpathogenic microorganisms as well as lifestyle can strongly affect a person's risk in developing neurodegenerative disorders either directly or indirectly.

2.3.1 Effect of Environmental Exposures: The In Utero Example

One such example is how the environment and exposures surrounding a fetus during pregnancy can actually affect its adult life (Gluckman et al. 2008). It has been proven that inadequate environmental conditions during pregnancy as well as infancy can strongly increase the risk of neurodegenerative disorders. The maternal environment can strongly impact the future of her child as during the in utero phase; the fetus is only exposed to the environment through its mother. This is even evident as the growth of the fetus is constrained by the mother's body size, stature, and parity. During pregnancy, a human fetus can be exposed to hypoxia, metals, hormones, and other substances (Charalambous et al. 2007). It has been proven that exposure to hypoxenima during pregnancy can lead to synaptic dysfunctions in the fetus, causing neurodegeneration (Morales et al. 2008). As for hormonal disturbances, in the case that the placental barrier is compromised, certain hormones such as glucocorticoids can travel through the placenta and affect the fetus, leading to altered programming of hypothalamus-pituitary-adrenal (HPA) (Kapoor et al. 2008). Human studies that follow the effect of environmental exposures on fetal development and neurodegeneration are limited by either their retrospective nature or difficulty of their implementation in the case of longitudinal studies. In animals, a model of prenatal asphyxia has demonstrated that retinal development is impaired due to the reduction of ganglion cells as a result of degeneration which is anticipated to lead to long-term complications (Piscopo et al. 2008). In another animal study on rabbits, placental insufficiency was directly associated with brain damage through the impairment of metabolic processes (van Vliet et al. 2013). Studies have repeatedly shown that environmental exposures on the developing brain have far more detrimental effects than those on the adult brain. In yet another interesting animal study, it was found that when the mother was subjected to lipopolysaccharide treatment during pregnancy, the pups turned out to suffer from loss of dopaminergic neurons, giving insight into how exposure to LPS during pregnancy can increase susceptibility to parkinsonism (Ling et al. 2002). In addition, exposure to metal toxins was found to induce altered levels of antioxidants in rats accompanied by increases in oxidative stress (Erikson et al. 2006). Collectively, these studies showcase how vital is to decipher the role of environmental exposures in utero in determining fetal as well as adult brain health.

2.4 The Gene-Environment Nexus

Research has shown that neurological diseases can be caused by the interaction between genetic and environmental factors. With the advancement in the arsenal of tools of genomic sciences, more evidence points to the fact that genetic modulators can strongly affect the risk and susceptibility to neurodegenerative diseases (Patel 2016). With these advancements, the goal now is to begin standardizing the definition of environmental exposure and how to analyze it. Newer types of studies include exposome studies that analyze the entirety of environmental exposures throughout the individual's lifespan (Wild 2005). Huge amounts of data are currently being generated both with regard to genetics and genomics as well as exposure; the major challenge remains, including how to study and integrate this data to begin elucidating how certain gene variants can control a human's response to environmental factors, hence altering their susceptibility to diseases including neurodegenerative ones (Rappaport 2011). This nexus of interaction may actually prove to be incredibly useful in the case of diseases of aging as the accumulation of a lifetime of exposures can strongly alter a populations risk overdeveloping a certain disease (Shulman et al. 2011). Two major examples of this are Alzheimer's disease and Parkinson's disease which have been proven to have strong genetic components yet can strongly be affected by certain environmental exposures during a person's life (Barnes and Yaffe 2011). The interaction between these factors, however, is yet to be elucidated.

The extrapolation of data and the identification of points of interaction between gene and environment are very challenging, and even when this task is done correctly, the challenge remains on identifying how this interaction affects this disease development and progression (Ahmad et al. 2013). For example, the effect of this interaction may actually be very small, and in order to be well observed, it would require expensive studies that integrate a large number of research subjects to acquire their genomic data. Even if such studies are implemented, the acquiring of exposure data especially retrospectively is yet another challenge that needs to be overcome. Retrospective data often suffers from poor quality and this is evident in environment-wide association studies whose challenges include incomplete or biased recall, various confounding variables, difficulty of ascertaining the exact timing and duration of exposure, and finally sampling issues. The challenging nature of such studies suffers from yet another layer of added complexity when we consider that the environmental factors and exposures themselves evolve as human population and industries evolve. Certain environmental factors that are very prominent nowadays were not prominent a few decades ago and vice versa. These challenges can be circumvented to a certain extent by performing animal studies under controlled conditions and the added benefit of the shortened life span of laboratory animals. However, such studies are preliminary at best and can only serve as pointers to potential interactions. These studies are useful when studying already identified variants which cannot allow for the de novo identification of human polymorphisms (Ermann and Glimcher 2012).

Another way to circumvent these challenges is genome-wide association and interaction studies that have robust analysis algorithms allowing for strong predictive power and relatively smaller sample sizes (C. Lin et al. 2015). Furthermore, a lot of interest is currently directed towards the observation of epigenetic changes occurring in response to environmental exposure as potential hints to gene-environment interactions (Maloney et al. 2012). Advances in technologies that can be used for tracking environmental exposure data are also strongly overcoming the challenges of inaccurate reporting, such technologies include wearables and smartphone technologies (Ueberham and Schlink 2018). The interest of governmental agencies in funding research in this area has allowed for the implementation of large-scale prospective studies that are currently providing large amounts of data both on the genetic and environment fronts, allowing for a true exploration of gene-environment interactions (McAllister et al. 2017).

Animal studies are also progressing by the enhancement of animal strains by diversifying their genetic makeup allowing for more complex studies under a controlled lab environment (Schoenrock et al. 2018; Williams et al. 2016). This is exceptionally useful in the area of gene environment interaction studies as mouse models allow for validation of findings found in humans as well as potential identification of novel interactions going hand in hand with large-scale human studies (Jones et al. 2013).

2.5 How to Approach Gene-Environment Interaction Studies in the Neurodegenerative Context

2.5.1 Approach 1: One Gene-One Exposure

In an attempt to elucidate whether certain disease variance can affect individual susceptibility to an environmental exposure, various studies have adopted the one gene-one exposure model (Fig. 2.1) (Dunn et al. 2019). In this type of studies, researchers compare between those who are variant positive and those who are negative and evaluate their odds ratios once exposed to a certain environmental factor. In these studies, it is possible that the accumulation of the exposure along

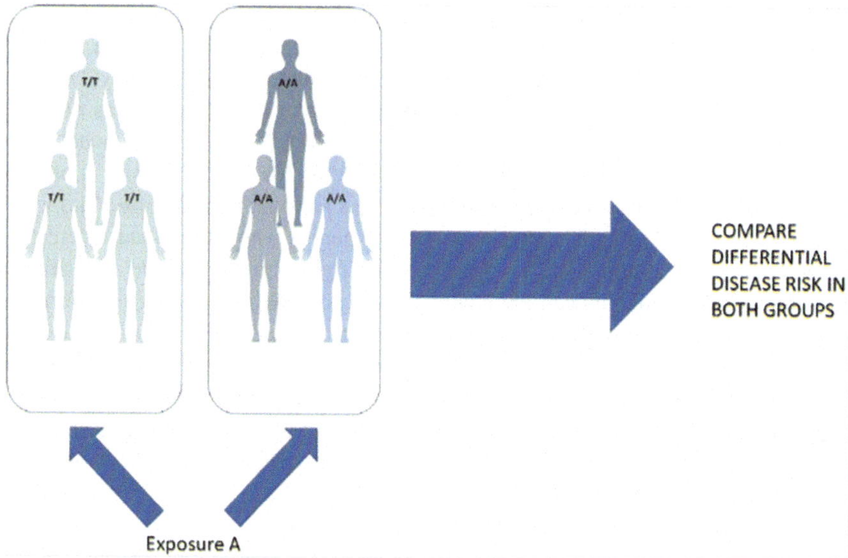

Fig. 2.1 Candidate gene analyses ask whether an exposure or exposures result in differential disease risk in carriers (e.g., "T/T" individuals) or noncarriers (e.g., "A/A" individuals) of a particular known disease risk allele (Dunn et al. 2019)

with the presence of the disease variant can both strongly lead to the development of neurodegenerative diseases (Lill 2016). One such example is studies evaluating the effect of dietary interventions in those carrying certain variants in the APOE gene on cholesterol levels and susceptibility to neurodegenerative diseases (Head et al. 2012). Both human and animal studies have shown that a person's APOE genotype affects their response to various environmental risk factors such as poor diet, obesity, stress, poor lifestyle, and heavy metal exposure by making them more susceptible to the development of neurodegenerative disorders (Pankratz and Foroud 2004). Collectively this evidence shows that certain risk factors that are genetic in nature can act together with environmental exposures and strongly affect this risk. Some other studies have shown that certain APOE genotypes can actually be protective against the effects of certain environmental factors (Wirth et al. 2014). One example is how APOE for carriers benefit from cognitive activity and are relatively protected from Alzheimer's disease one compared to APOE 2 and 3 carriers (Rajan et al. 2014).

2.5.2 *Approach 2: Gene Datasets—Multiple Exposures*

The complexity of the human genome and environmental exposures guarantees that the one gene-one exposure approach is potentially faulty and may not be the most accurate way to begin inferring relevant gene-environment interactions. Several studies have attempted to go beyond this model as an intermediary approach between the previous one and genome-wide associations. For example, a study observing the effect of a multitude of lifestyle choices on a panel on 27 genes associated with Alzheimer's found that one genotype (SLC24A4) was strongly affected by whether the person was a smoker or a drinker and in turn impacted the person's risk of developing the disease (E. Lin et al. 2017). In a similar approach, another group looked at the relationship between cardiovascular disease risk factors and another panel of Alzheimer's-associated genes. They found an interesting correlation between the ABCA7 genotype and right parietal volume indicating a potential relationship between this genotype and susceptibility to cognitive impairment induced by cardiovascular-associated factors (Wang et al. 2017). This type of studies does offer significant advantages to those conferred by one gene-one exposure-type studies as it eliminates certain biases and provides insights as to which pathways may be affected by environmental exposures, thus mediating disease. Yet, despite their usefulness, they do not enable de novo identification of relevant genes/ pathways (Holmans et al. 2013).

2.5.3 Approach 3: Observing the Genome

As previously discussed, approaches that start with a pre-identified list of candidate genes and exposures is minimally useful in the de novo identification of gene-environment interactions that affect disease susceptibility. In the case of mapping environmental exposure, the best study design remains to be genome-wide association and interaction studies (GWAIS) (Fig. 2.2). When conducted in the realm of neurobiology, these studies have strikingly identified risk genes and exposures that

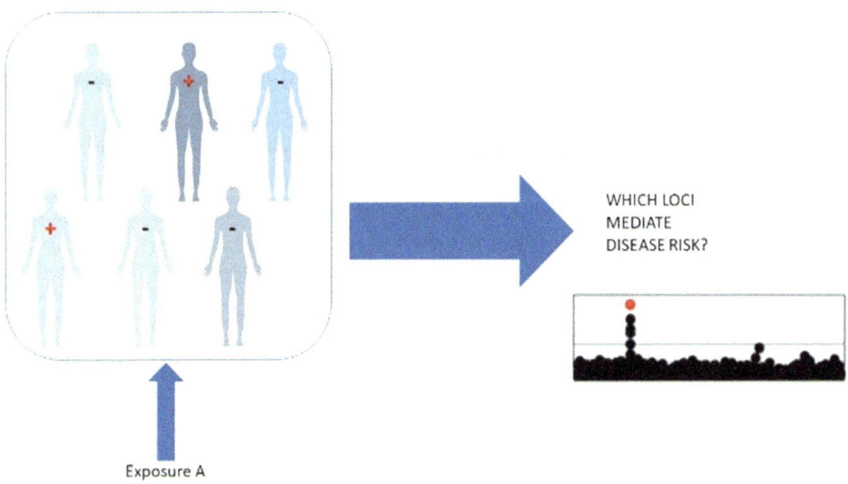

Fig. 2.2 GWAIS ask which genetic variants and genomic loci correlate with disease risk given individuals' exposure to a known disease-relevant environmental factor (Dunn et al. 2019)

were previously implicated in diseases of different nature, hinting at the importance of expanding neurobiology research even beyond the genome-wide model. GWAIS have identified interactions between novel genetic modulators and environmental exposures in Alzheimer's and Parkinson's (W.-Y. Lin et al. 2019). For example, one of these studies found that glutamate receptor genes and their variants are deterministic factors in a person's response to caffeine in Parkinson's disease. One interesting variant was the rs4998386 SNP of GRIN2A which was found to be strongly associated with a 70–80% caffeine-induced reduction in Parkinson's Disease risk of development (Hamza et al. 2011). These findings were identical in two studies and irreproducible in a third, raising the doubt that this relationship may count on another unidentified factor, further adding to the complexity of the gene-environment interaction in this scenario (Ahmed et al. 2014; Yamada-Fowler et al. 2014). Another GWAIS approach shed light on the relationship between smoking and the risk of Parkinson's showing that according to the individual genetic landscape, smoking may actually confer protective properties from the disease. Implicated genes included variants of gene known as the synaptic vesicle glycoprotein 2 (SV2C) gene. Homozygous major allele individuals showed a reduction in Parkinson's disease risk associated with smoking, while those homozygous for the minor alleles showed an increased risk induced by smoking (Hill-Burns et al. 2013). These results were also validated in animal studies where SV2C knockout mice showed drastically different dopamine responses to nicotine than wild-types (Dunn et al. 2017). Despite the promise of GWAIS-type studies, our understanding of the gene-environment nexus will also remain predominantly hindered by the incomplete nature of exposure data in human studies.

2.6 What the Future Holds for the Gene-Environment Nexus in Neurodegenerative Disorders

Evidence has repeatedly shown that genetics have a strong bearing on the risk and development of neurodegenerative diseases and so does environmental exposure. Furthermore, evidence strongly suggests that certain genetic variants modulate a person's response to environmental exposures and have the capacity of greatly affecting their susceptibility to disease and its prognosis. GWAIS approaches are currently under constant refinement to allow for expanded genome-wide investigations of gene-environment interactions. However, contrary to common belief, the real challenge lies in the tracking and reporting of exposure data that does not suffer from the effects of recall bias and inaccurate reporting.

The apparent solution is large-scale prospective longitudinal studies where the researchers collect exposure data in a timely and accurate manner (Fig. 2.3). Indeed, various studies of this nature have been initiated such as the UK Biobank study which is collecting general exposure data and genomic sequences of approximately 50,000 participants (Sudlow et al. 2015). A similar study is found in the USA under the name "All of Us," an NIH-funded program collecting massive amounts of

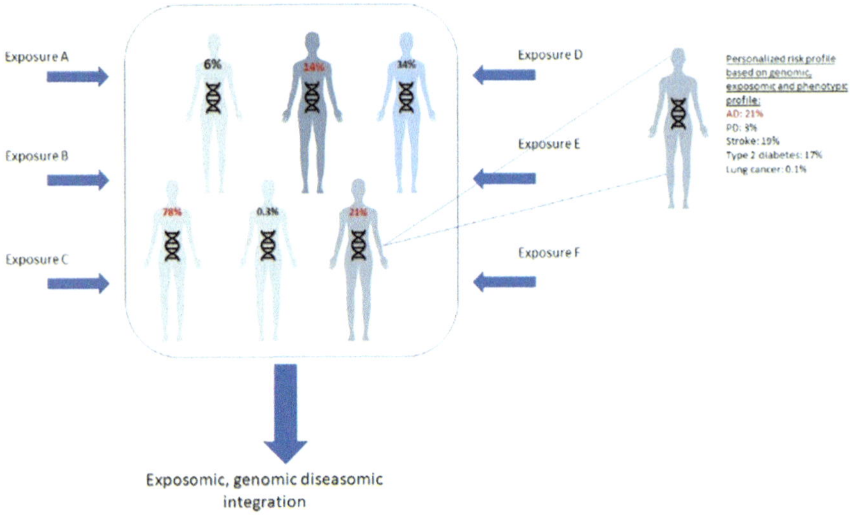

Fig. 2.3 GxEWAS take an integrated approach of measuring and determining which exposome-wide and genome-wide factors contribute to the risk of a variety of diseases and how these factors interact across time in order to determine individual risk for any number of diseases (Dunn et al. 2019)

genomic, environmental, and health data from a million participants (Lyles et al. 2018). These studies are achieving what previous studies couldn't in terms of the sheer amounts of data being collected as well as the diversity of their participants and sample size. Furthermore, they are heavily backed with the most recent advances in analysis tools in the field of bioinformatics (Bradley et al. 2018).

It is inevitable, however, that these studies will be affected by some of the hurdles that affected their predecessors like selection and survival bias or attrition or different confounding variables. This is complemented by the ongoing research and enhancement in the quality of animal studies as a means of evaluating and validating the associations revealed by these studies. Furthermore, a multitude of other factors needs to be considered including duration and age at exposure (C. Lin et al. 2015). The data to be collected from these large-scale studies as well as other cohorts of both human and animal studies are bound to provide valuable insights with regard to the role of gene-environment interactions in neurodegenerative diseases.

References

Ahmad S, Rukh G, Varga TV, Ali A, Kurbasic A, Shungin D, Ericson U, Koivula RW, Chu AY, Rose LM, Ganna A, Qi Q, Stančáková A, Sandholt CH, Elks CE, Curhan G, Jensen MK, Tamimi RM, Allin KH et al (2013) Gene × physical activity interactions in obesity: combined analysis of 111,421 individuals of European ancestry. PLoS Genet 9(7):e1003607. https://doi.org/10.1371/journal.pgen.1003607

Ahmed I, Lee P-C, Lill CM, Nielsen SS, Artaud F, Gallagher LG, Loriot M-A, Mulot C, Nacfer M, Liu T, Biernacka JM, Armasu S, Anderson K, Farin FM, Lassen CF, Hansen J, Olsen JH, Bertram L, Maraganore DM et al (2014) Lack of replication of the GRIN2A-by-coffee interaction in Parkinson disease. PLoS Genet 10(11):e1004788. https://doi.org/10.1371/journal.pgen.1004788

Alkanli N, Ay A (2019) The relationship between alpha-synuclein (SNCA) gene polymorphisms and development risk of Parkinson's disease. In: Synucleins—biochemistry and role in diseases. https://doi.org/10.5772/intechopen.82808

Barnes DE, Yaffe K (2011) The projected effect of risk factor reduction on Alzheimer's disease prevalence. Lancet Neurol 10(9):819–828. https://doi.org/10.1016/S1474-4422(11)70072-2

Bates GP (2005) The molecular genetics of Huntington disease—a history. Nat Rev Genet 6(10): 766–773. https://doi.org/10.1038/nrg1686

Bekris LM, Yu C-E, Bird TD, Tsuang DW (2010) Genetics of Alzheimer disease. J Geriatr Psychiatry Neurol 23(4):213–227. https://doi.org/10.1177/0891988710383571

Bradley WG, Andrew AS, Traynor BJ, Chiò A, Butt TH, Stommel EW (2018) Gene-environment-time interactions in neurodegenerative diseases: hypotheses and research approaches. Ann Neurosci 25(4):261–267. https://doi.org/10.1159/000495321

Brown RC, Lockwood AH, Sonawane BR (2005) Neurodegenerative diseases: an overview of environmental risk factors. Environ Health Perspect 113(9):1250–1256. https://doi.org/10.1289/ehp.7567

Charalambous M, da Rocha ST, Ferguson-Smith AC (2007) Genomic imprinting, growth control and the allocation of nutritional resources: consequences for postnatal life. Curr Opin Endocrinol Diabetes Obes 14(1):3–12. https://doi.org/10.1097/MED.0b013e328013daa2

Dunn AR, Stout KA, Ozawa M, Lohr KM, Hoffman CA, Bernstein AI, Li Y, Wang M, Sgobio C, Sastry N, Cai H, Caudle WM, Miller GW (2017) Synaptic vesicle glycoprotein 2C (SV2C) modulates dopamine release and is disrupted in Parkinson disease. Proc Natl Acad Sci U S A 114(11):E2253–E2262. https://doi.org/10.1073/pnas.1616892114

Dunn AR, O'Connell KMS, Kaczorowski CC (2019) Gene-by-environment interactions in Alzheimer's disease and Parkinson's disease. Neurosci Biobehav Rev 103:73–80. https://doi.org/10.1016/j.neubiorev.2019.06.018

Erikson KM, Dorman DC, Fitsanakis V, Lash LH, Aschner M (2006) Alterations of oxidative stress biomarkers due to in utero and neonatal exposures of airborne manganese. Biol Trace Elem Res 111(1–3):199–215. https://doi.org/10.1385/BTER:111:1:199

Ermann J, Glimcher LH (2012) After GWAS: mice to the rescue? Curr Opin Immunol 24(5): 564–570. https://doi.org/10.1016/j.coi.2012.09.005

Gatz M, Pedersen NL (2013) Study of dementia in Swedish twins. Twin Res Hum Genet 16(1). https://doi.org/10.1017/thg.2012.68

Gluckman PD, Hanson MA, Cooper C, Thornburg KL (2008) Effect of in utero and early-life conditions on adult health and disease. N Engl J Med 359(1):61–73. https://doi.org/10.1056/NEJMra0708473

Gusella JF, Wexler NS, Conneally PM, Naylor SL, Anderson MA, Tanzi RE, Watkins PC, Ottina K, Wallace MR, Sakaguchi AY, Young AB, Shoulson I, Bonilla E, Martin JB (1983) A polymorphic DNA marker genetically linked to Huntington's disease. Nature 306(5940): 234–238. https://doi.org/10.1038/306234a0

Hamza TH, Chen H, Hill-Burns EM, Rhodes SL, Montimurro J, Kay DM, Tenesa A, Kusel VI, Sheehan P, Eaaswarkhanth M, Yearout D, Samii A, Roberts JW, Agarwal P, Bordelon Y, Park Y, Wang L, Gao J, Vance JM et al (2011) Genome-wide gene-environment study identifies glutamate receptor gene GRIN2A as a Parkinson's disease modifier gene via interaction with coffee. PLoS Genet 7(8):e1002237. https://doi.org/10.1371/journal.pgen.1002237

Hannan AJ (2004) Molecular mediators, environmental modulators and experience-dependent synaptic dysfunction in Huntington's disease. Acta Biochim Pol 51(2):415–430. https://doi.org/035001415

Head D, Bugg JM, Goate AM, Fagan AM, Mintun MA, Benzinger T, Holtzman DM, Morris JC (2012) Exercise engagement as a moderator of the effects of APOE genotype on amyloid deposition. Arch Neurol 69(5):636–643. https://doi.org/10.1001/archneurol.2011.845
Hill-Burns EM, Singh N, Ganguly P, Hamza TH, Montimurro J, Kay DM, Yearout D, Sheehan P, Frodey K, Mclear JA, Feany MB, Hanes SD, Wolfgang WJ, Zabetian CP, Factor SA, Payami H (2013) A genetic basis for the variable effect of smoking/nicotine on Parkinson's disease. Pharmacogenomics J 13(6):530–537. https://doi.org/10.1038/tpj.2012.38
Hirschhorn JN, Daly MJ (2005) Genome-wide association studies for common diseases and complex traits. Nat Rev Genet 6(2):95–108. https://doi.org/10.1038/nrg1521
Holmans P, Moskvina V, Jones L, Sharma M, The International Parkinson's Disease Genomics Consortium (IPDGC), Vedernikov A, Buchel F, Sadd M, Bras JM, Bettella F, Nicolaou N, Simón-Sánchez J, Mittag F, Gibbs JR, Schulte C, Durr A, Guerreiro R, Hernandez D, Brice A et al (2013) A pathway-based analysis provides additional support for an immune-related genetic susceptibility to Parkinson's disease. Hum Mol Genet 22(5):1039–1049. https://doi.org/10.1093/hmg/dds492
Jones BC, Miller DB, O'Callaghan JP, Lu L, Unger EL, Alam G, Williams RW (2013) Systems analysis of genetic variation in MPTP neurotoxicity in mice. Neurotoxicology 37:26–34. https://doi.org/10.1016/j.neuro.2013.03.010
Kapoor A, Petropoulos S, Matthews SG (2008) Fetal programming of hypothalamic-pituitary-adrenal (HPA) axis function and behavior by synthetic glucocorticoids. Brain Res Rev 57(2): 586–595. https://doi.org/10.1016/j.brainresrev.2007.06.013
Klein C, Westenberger A (2012) Genetics of Parkinson's disease. Cold Spring Harb Perspect Med 2(1). https://doi.org/10.1101/cshperspect.a008888
Lill CM (2016) Genetics of Parkinson's disease. Mol Cell Probes 30(6):386–396. https://doi.org/10.1016/j.mcp.2016.11.001
Lin C, Chu C-M, Lin J, Yang H-Y, Su S-L (2015) Gene-gene and gene-environment interactions in meta-analysis of genetic association studies. PLoS One 10(4):e0124967. https://doi.org/10.1371/journal.pone.0124967
Lin E, Tsai S-J, Kuo P-H, Liu Y-L, Yang AC, Kao C-F (2017) Association and interaction effects of Alzheimer's disease-associated genes and lifestyle on cognitive aging in older adults in a Taiwanese population. Oncotarget 8(15):24077–24087. https://doi.org/10.18632/oncotarget.15269
Lin W-Y, Huang C-C, Liu Y-L, Tsai S-J, Kuo P-H (2019) Genome-wide gene-environment interaction analysis using set-based association tests. Front Genet 9. https://doi.org/10.3389/fgene.2018.00715
Ling Z, Gayle DA, Ma SY, Lipton JW, Tong CW, Hong J-S, Carvey PM (2002) In utero bacterial endotoxin exposure causes loss of tyrosine hydroxylase neurons in the postnatal rat midbrain. Mov Disord 17(1):116–124. https://doi.org/10.1002/mds.10078
Logroscino G, Tortelli R (2015) Epidemiology of neurodegenerative diseases. In: Imaging in neurodegenerative disorders. Oxford University Press. https://oxfordmedicine.com/view/10.1093/med/9780199671618.001.0001/med-9780199671618-chapter-1
Lyles CR, Lunn MR, Obedin-Maliver J, Bibbins-Domingo K (2018) The new era of precision population health: insights for the all of us research program and beyond. J Transl Med 16(1): 211. https://doi.org/10.1186/s12967-018-1585-5
Maloney B, Sambamurti K, Zawia N, Lahiri DK (2012) Applying epigenetics to Alzheimer's disease via the latent early–life associated regulation (LEARn) model. Curr Alzheimer Res 9(5): 589–599. https://www.eurekaselect.com/97893/article
Mayeux R (2003) Epidemiology of neurodegeneration. Annu Rev Neurosci 26:81–104. https://doi.org/10.1146/annurev.neuro.26.043002.094919
McAllister K, Mechanic LE, Amos C, Aschard H, Blair IA, Chatterjee N, Conti D, Gauderman WJ, Hsu L, Hutter CM, Jankowska MM, Kerr J, Kraft P, Montgomery SB, Mukherjee B, Papanicolaou GJ, Patel CJ, Ritchie MD, Ritz BR et al (2017) Current challenges and new

opportunities for gene-environment interaction studies of complex diseases. Am J Epidemiol 186(7):753–761. https://doi.org/10.1093/aje/kwx227

Minghetti L (2005) Role of inflammation in neurodegenerative diseases. Curr Opin Neurol 18(3): 315–321. https://doi.org/10.1097/01.wco.0000169752.54191.97

Morales P, Fiedler JL, Andrés S, Berrios C, Huaiquín P, Bustamante D, Cardenas S, Parra E, Herrera-Marschitz M (2008) Plasticity of hippocampus following perinatal asphyxia: effects on postnatal apoptosis and neurogenesis. J Neurosci Res 86(12):2650–2662. https://doi.org/10.1002/jnr.21715

Pankratz N, Foroud T (2004) Genetics of Parkinson disease. NeuroRx 1(2):235–242

Patel CJ (2016) Analytical complexity in detection of gene variant-by-environment exposure interactions in high-throughput genomic and exposomic research. Curr Environ Health Rep 3(1):64–72. https://doi.org/10.1007/s40572-016-0080-5

Pihlstrøm L, Wiethoff S, Houlden H (2018) Chapter 22—genetics of neurodegenerative diseases: an overview. In: Kovacs GG, Alafuzoff I (eds) Handbook of clinical neurology, vol 145. Elsevier, pp 309–323. https://doi.org/10.1016/B978-0-12-802395-2.00022-5

Pilotto A, Padovani A, Borroni B (2013) Clinical, biological, and imaging features of monogenic Alzheimer's disease. Biomed Res Int 2013. https://doi.org/10.1155/2013/689591

Piscopo P, Bernardo A, Calamandrei G, Venerosi A, Valanzano A, Bianchi D, Confaloni A, Minghetti L (2008) Altered expression of cyclooxygenase-2, presenilins and oxygen radical scavenging enzymes in a rat model of global perinatal asphyxia. Exp Neurol 209(1):192–198. https://doi.org/10.1016/j.expneurol.2007.09.014

Rajan KB, Skarupski KA, Rasmussen HE, Evans DA (2014) Gene-environment interaction of body mass index and apolipoprotein E ε4 allele on cognitive decline. Alzheimer Dis Assoc Disord 28(2):134–140. https://doi.org/10.1097/WAD.0000000000000013

Rappaport SM (2011) Implications of the exposome for exposure science. J Expo Sci Environ Epidemiol 21(1):5–9. https://doi.org/10.1038/jes.2010.50

Reddy PH (2009) The role of mitochondria in neurodegenerative diseases: mitochondria as a therapeutic target in Alzheimer's disease. CNS Spectr 14(8 Suppl 7):8–18

Schoenrock SA, Oreper D, Farrington J, McMullan RC, Ervin R, Miller DR, Pardo-Manuel de Villena F, Valdar W, Tarantino LM (2018) Perinatal nutrition interacts with genetic background to alter behavior in a parent-of-origin-dependent manner in adult collaborative cross mice. Genes Brain Behav 17(7):e12438. https://doi.org/10.1111/gbb.12438

Shulman JM, De Jager PL, Feany MB (2011) Parkinson's disease: genetics and pathogenesis. Annu Rev Pathol 6(1):193–222. https://doi.org/10.1146/annurev-pathol-011110-130242

Sudlow C, Gallacher J, Allen N, Beral V, Burton P, Danesh J, Downey P, Elliott P, Green J, Landray M, Liu B, Matthews P, Ong G, Pell J, Silman A, Young A, Sprosen T, Peakman T, Collins R (2015) UK biobank: an open access resource for identifying the causes of a wide range of complex diseases of middle and old age. PLoS Med 12(3):e1001779. https://doi.org/10.1371/journal.pmed.1001779

Ueberham M, Schlink U (2018) Wearable sensors for multifactorial personal exposure measurements—a ranking study. Environ Int 121:130–138. https://doi.org/10.1016/j.envint.2018.08.057

van Vliet E, Eixarch E, Illa M, Arbat-Plana A, González-Tendero A, Hogberg HT, Zhao L, Hartung T, Gratacos E (2013) Metabolomics reveals metabolic alterations by intrauterine growth restriction in the Fetal rabbit brain. PLoS One 8(5):e64545. https://doi.org/10.1371/journal.pone.0064545

Wang C, Sun J, Guillaume B, Ge T, Hibar DP, Greenwood CMT, Qiu A, Initiative, The A. D. N (2017) A set-based mixed effect model for gene-environment interaction and its application to neuroimaging phenotypes. Front Neurosci 11. https://doi.org/10.3389/fnins.2017.00191

Wild CP (2005) Complementing the genome with an "exposome": the outstanding challenge of environmental exposure measurement in molecular epidemiology. Cancer Epidemiol Prev Biomark 14(8):1847–1850. https://doi.org/10.1158/1055-9965.EPI-05-0456

Williams EG, Wu Y, Jha P, Dubuis S, Blattmann P, Argmann CA, Houten SM, Amariuta T, Wolski W, Zamboni N, Aebersold R, Auwerx J (2016) Systems proteomics of liver mitochondria function. Science (New York, NY) 352(6291):aad0189. https://doi.org/10.1126/science.aad0189

Wirth M, Villeneuve S, Joie RL, Marks SM, Jagust WJ (2014) Gene–environment interactions: lifetime cognitive activity, APOE genotype, and beta-amyloid burden. J Neurosci 34(25): 8612–8617. https://doi.org/10.1523/JNEUROSCI.4612-13.2014

Yamada-Fowler N, Fredrikson M, Söderkvist P (2014) Caffeine interaction with glutamate receptor gene GRIN2A: Parkinson's disease in Swedish population. PLoS One 9(6):e99294. https://doi.org/10.1371/journal.pone.0099294

Chapter 3
Gene-Gut-Brain Axis: Gene-Based Personalized Medicine

Samir A. Salama and Duaa Dakhlallah

3.1 Introduction

The human genome is composed of three billion base pairs that include coding and noncoding sequences. The coding sequences encode for more than 20,000 genes that are responsible for different human phenotypes (Salzberg 2018). While the primary sequence of nucleotide in the human genome dictates the individual's traits, modification of DNA nucleotide or the histone proteins has profound effects on the gene expression pattern and the resulting phenotypes. Such modifications are known as epigenetic modifications which substantially affects the individual's phenotypic characteristics without changing the primary structure of the DNA (Yi and Goodisman 2021). A study of identical twins revealed that interaction between the genome and the environment considerably affects the phenotypic outcome irrespective of the similarity of the primary sequence of the DNA (Fraga et al. 2005). While environmental factors can affect the individual's phenotype through changing the primary sequence of the DNA (induction of mutations), modification of the epigenome has a significant impact as a mediator of environmental effects on the individual's phenotypic characteristics. Among the environmental factors that modulate the epigenome, nutrition has a great influence (Mullins et al. 2020). A study of the interaction between the nutrition and individual's genome in order to guide

S. A. Salama
Department of Biochemistry, Faculty of Pharmacy, Al-Azhar University, Cairo, Egypt

D. Dakhlallah (✉)
Institute of Global Health and Human Ecology, School of Sciences and Engineering, The American University of Cairo, New Cairo, Egypt

Microbiology, Immunology and Cell Biology, and Member of Mary Babb Randolph Cancer Center, Erma Byrd Biomedical Research Center, West Virginia University, Morgantown, WV, USA
e-mail: duaa.dakhlallah@aucegypt.edu

M. Salama (ed.), *Nutrigenomics and the Brain*, Nutritional Neurosciences,
https://doi.org/10.1007/978-981-16-9205-5_3

individually tailored nutritional intervention is a promising field not only for diseases prevention and management but also for health improvement (Astley 2007; Peregrin 2001). In the future, personalized nutritional advice will be easily suggested based on the individual genetic variation by the aid of the advanced "omics" approaches.

3.2 Nutrigenomics: How Nutrients Affect Genes

Nutrients can affect the gene expression pattern and consequently the phenotypic characteristics of individuals (Müller and Kersten 2003). Nutrigenomics is the branch of science that deals with the ability of nutrients to modify the gene expression aiming at identifying health-related effects of nutrients (Miggiano and De Sanctis 2006; Remely et al. 2015a). Determining the relation between nutrition patterns and the development of certain diseases, especially the chronic disease, supports the prevention or delay of such diseases (Miggiano and De Sanctis 2006). The ultimate goal of nutrigenomics is, thus, to guide individualized nutritional intervention in order to help treating, modifying, or preventing diseases as well as improving the health of human beings. Epigenetics represent an important link that connects many environmental factors, including nutrition, to the genome. Nutrients can modify the epigenome through different mechanisms. They can supply the required material for DNA methylation or histone modifications. Additionally, they can modulate the enzymatic activity of DNA methyltransferases or one carbon metabolism-related enzymes (Remely et al. 2015a; Zhang 2015). Such modifications modulate the accessibility of gene promoter regions by the transcription factors, affecting the gene expression (Kaput et al. 2007). While the mechanisms mediate the ability of some nutrient to modulate the epigenome have been uncovered, a lot of others still to be addressed.

3.3 Nutrients Modulate DNA Methylation and Histone Acetylation

DNA methylation is the most stable epigenetic modification of the genome (Bordoni and Gabbianelli 2019). Methyl groups are added to the cytosine nucleotide at carbon 5 in the CpG (cytosine-guanine)-rich regions in the DNA. Adding such methyl groups suppresses the gene expression which directly affects several biological outcomes, including cell proliferation and aging (Alam et al. 2019; Jones and Laird 1999). Micronutrients such as vitamin B12, vitamin B6, folate, and methionine which contribute to the one carbon metabolism play critical roles in DNA methylation process (Kaput et al. 2007). They increase the availability of the methyl donor s-adenosyl methionine. Defective methylation of DNA increases susceptibility to cancer and neural tube defects (Stover and Garza 2002). Other nutrients including retinoic acid, resveratrol, and curcumin have been implicated in the modification of DNA methylation (Alam et al. 2019; Shu et al. 2011). In addition to the role of micronutrients, caloric restriction plays important roles in DNA methylation. It has been demonstrated that caloric restriction increases methylation of RAS oncogene (Hass et al. 1993).

Besides DNA methylation, histone acetylation represents an important epigenetic mechanism that controls gene expression through altering the accessibility of the DNA by the transcription factors. Caloric control and NADH/NAD$^+$ ratio play important roles in the histone modifications (Blander and Guarente 2004). Caloric control has been associated with SIRT1 (NAD$^+$-dependent histone deacetylase) and HDAC1 gene expressions which stimulate histone deacetylation and induce expression of a variety of important genes including the tumor suppressor p53 gene (Alam et al. 2019; Blander and Guarente 2004). Permanent epigenetic modifications may be caused by the continuous usage of nutrient that induces such epigenetic modifications (Kaput et al. 2007).

3.4 Genetic Variations Affect the Response to Nutrients

In contrast to the effect of nutrients on the genome, genetic variation between individuals affects their response to the same nutrient. How the genetic makeup of individuals affects their response to the same nutrients is the subject of the nutrigenetics branch of science (German 2005; Hawkinson 2007). Inborn errors of metabolism exemplify how genetic variations affect the individual response to nutrition. Phenylketonuria induces mental retardation due to abnormal metabolic manipulation of the dietary phenylalanine amino acid. Similarly, deficiency in the lactase enzyme causes lactose intolerance (Gaboon 2011). Maple syrup urine disease is another example of the genetic variation that affects the metabolism of the branched-chain amino acids (Bordoni and Gabbianelli 2019; Kohlmeier 2012). Mutation in homogentisate dioxygenase gene results in alkaptonuria (Garrod 1902). While monogenetic variation is responsible for the abovementioned examples, more complex ploygenetic variations that are responsible for diseases such as obesity also exist (Fu et al. 2015; Shungin et al. 2015).

The human genome is almost the same in all individuals. Despite the obvious phenotypic variability between individuals, the genetic makeup of any two individuals differs only in less than 1% (Mullins et al. 2020). Such variations occur mostly in the noncoding regions of the genome which represents about 99% of the entire genome. The most common genetic variation between individual is the single nucleotide polymorphism (SNP) in which individuals differ in only one nucleotide at a certain DNA locus. The human genome contains about 300,000 SNPs that is about 1 SNP for every 1000 base pair (Nelson et al. 2004). The current researches focus on the relation of SNP variability and nutrition although other types of genetic variation such as insertion/deletions/inversion may also have roles (Mullins et al. 2020). Genome-wide genetic association studies are used to associate certain SNPs with certain phenotypic outcomes, but studies that are performed on certain population cannot reliably be applied to other populations (Medina-Gomez et al. 2015). DNA microarray is a widely used technique for these association studies to detect a huge number of SNPs (Hoffmann et al. 2011). While the majority of individual SNPs have no diet-related effects, others have a significant impact on nutrient-gene interaction. Common example of these SNPs include rs762551 in CYP1A2 gene that affects the individual ability to metabolize caffeine, rs1229984 and rs2066702 in ADH1B gene that affect the alcohol metabolism, and rs738409 in PNPLA3 that affects fat accumulation in the liver and the susceptibility to the fatty liver disease (Edenberg 2007; Frary et al. 2005; Mazo et al. 2019; Nehlig 2018; Sachse et al. 2003). Other examples include rs9939609 in FTO gene that affects the susceptibility to obesity and rs7412 and rs429358 in APOE gene that affect susceptibility to cardiovascular diseases and Alzheimer's disease, respectively (Ağagündüz and Gezmen-Karadağ 2019; Di Renzo et al. 2019; Duicu et al. 2016; Fawcett and Barroso 2010; Hardy et al. 2020; Kühn et al. 2016; Liu et al. 2019; Martins et al. 2006). Additionally, rs1801133 in MTHFR gene affects the folate metabolism, and rs7041 and rs4588 in GC gene affect the transport of vitamin D. While single SNP variability may underline a variety of nutrition-related problems, many nutrition-related problems are multifactorial and involve several genetic variations (Mullins et al. 2020). Identifying such SNPs is currently challenging. Commercially available genetic testing has been recently expanded to test for the common nutrition-related genetic variation in order to guide individually tailored nutritional advice, but its accuracy is still to be improved (Caulfield et al. 2010).

3.5 Nutrigenomics and Control of Chronic Diseases

Chronic diseases represent a leading cause of death in the United States although the majority of them can be prevented (Control, C.F.D. and Prevention 2009). The prevalence of chronic diseases such as cardiometabolic diseases and cancer has been largely attributed to nutrition problems (Afshin et al. 2019; Micha et al. 2017). Diet represents an important modifiable factor that has profound effects on the heath outcome. Consumption of certain types of diet such as fruits, vegetables, nuts, and fiber-rich diets decreases the risk of the cardiometabolic disease and diabetes mellitus (Qian et al. 2019; Satija and Hu 2018). Studies have heighted the role of fruit and vegetable consumption in controlling not only cancer but also the cardiovascular diseases, diabetes, and aging (Atanasov et al. 2015; Braicu et al. 2017). A variety of secondary metabolites in plants (phytochemicals) are responsible for the observed modulatory activity (Milenkovic et al. 2012; Waltenberger et al. 2016). For example, curcumin, a phytochemical, reduces the incidence of type 2 diabetes in the prediabetic subjects (Chuengsamarn et al. 2012). Other phytochemicals including resveratrol and epigallocatechin are implicated in the control of body weight and obesity (Milagro et al. 2013). Biologically active compounds derived from olive oil including oleic acid and biophenols have been shown to play important role in cancer chemoprevention (Braicu et al. 2017; Piroddi et al. 2017). The effect of these compounds is thought to be through the epigenetic modulatory activity (Remely et al. 2015b). Understanding the nutrient-gene-disease interconnection will help in developing strategies for individualized intervention to prevent and control diseases as well as improving health.

3.6 Nutrigenomic in Cardiovascular Diseases

Cardiovascular diseases (CVD) are worldwide prevalent diseases. Despite their complicated multifactorial etiology, CVD are extensively studied for understanding their genetic background (Corella and Ordovas 2009). Nutritional intervention has been explored not only to treat but also to prevent the CVD (Corella and Ordovas 2009; Reen et al. 2015). Olive oil consumption has been associated with improved cardiovascular function, decreased blood pressure, and improved lipid profile (De Santis et al. 2019). The polyphenolic content of olive oil has been implicated in modulation of inflammatory pathways, cellular redox status, and lipid metabolism (De Santis et al. 2019). Oxidative modification of low-density lipoproteins (LDL) has been significantly diminished with administration of olive oil for 3 weeks (Castaner et al. 2012). The underlining molecular mechanism was found to be reduced CD40/CD40-ligand interaction with subsequent reduction in the inflammatory responses and pro-inflammatory cytokine levels (Castaner et al. 2012). Polyphenolic content of the olive oil reduces gene expression of IL8RA that modulates the blood pressure through targeting the renin-angiotensin-aldosterone system (De Santis et al. 2019). Administration of olive oil with high polyphenolic content also suppresses expression of the inflammation-related genes IFN and IL-7R and the oxidative stress-related gene ADRB2 (Konstantinidou et al. 2010). Omega-3 polyunsaturated fatty acids have been shown to reduce the release of leukotriene that is increased at the atherosclerotic sites through modulation of lipoxogenase-5 gene expression (Kaur et al. 2018). Yoo, J. and Park, S. demonstrated that the interaction between low folate levels and the *MTHFR* gene with C677T SNP determines the risk level of the coronary artery disease (Yoo and Park 2000). Vitamin B interacts with the same SNP to determine the level of homocysteine and subsequently the risk of thromboembolic vulnerability (Yates and Lucock 2003). Yang Y. and colleagues showed that dietary saturated fats interact with E2 and E4 SNP of APOE gene to increase vulnerability to myocardial infarction (Yang et al. 2007). Dietary arachidonic acid interacts with SNPs in 5-lipoxygenase gene and increases the risk of myocardial infarction (Allayee et al. 2008).

3.7 Nutrigenomics in Cancer

Cancer is an uncontrolled cell division with multifactorial etiology. Diet-derived compounds have been demonstrated to modulate different stages of cancer including angiogenesis and metastasis (Braicu et al. 2017). It has been reported that more than a quarter of cancer types can be modulated with diet (Ardekani and Jabbari 2009). Diet rich in fish and green tea has been associated with decreased rate of breast cancer in Asian population (Petric et al. 2015). Phytochemicals are plant-derived secondary metabolites that are produced to defend plants or to provide attractive color or smell (Liu 2004). These phytochemicals can be classified into sulfhydryl compounds, polyphenols, or terpenoids. Among these compounds, quercetin and Kaempferol (flavonols), luteolin and apigenin (flavones), genistein and daidzein (isoflavones), and naringenin and hesperidin (flavanones) have been extensively studied (Braicu et al. 2017; Hardman 2014). Flavones and isolflavones represent plant-derived estrogen-like compounds that interfere with estrogen-related molecular pathways (He and Chen 2013). The isoflavones such as genistein and daidzein have also been demonstrated to control breast cancer and prostate cancer via targeting estrogenic and androgenic receptors, respectively (Adjakly et al. 2013; Choi and Kim 2013; Upadhyay and Dixit 2015). Phytochemicals also have been shown to modulate genomic-related molecular pathways in lung, breast, colon, liver, ovarian, and prostate cancers (Del Follo-Martinez et al. 2013; Du et al. 2013; Hua et al. 2016; Kim et al. 2013; Omene et al. 2013; Ozturk et al. 2012; Singh et al. 2017; Tolba et al. 2013; Weng and Yen 2012; Wubetu et al. 2016). Epigallocatechin-3-gallate, a major polyphenolic compound in green tea, has been implicated in epigenetic regulation of gene expression in cancer cells through histone modification (Meeran et al. 2011; Nandakumar et al. 2011). Apigenin has been shown to downregulate DNMT and HDAC activity in the skin epidermal cells (Paredes-Gonzalez et al. 2014). Genistein plays role in DNA methylation and histone acetylation and activates the tumor suppressor genes (Kikuno et al. 2008; Vardi et al. 2010). These phytochemical compounds have been implicated in modulation of a large variety of cancer-related genes (Braicu et al. 2017).

3.8 Conclusions and Future Directions

Nutrition represents a vital environmental factor that affects our health. Mutual interaction between nutrition and our genes largely determine the individual health outcome. Genes affect the metabolism of nutrition, while nutrients affect gene expression patterns. Switching from the conventional nutrition guideline (i.e., based on factors such as gender or age groups) to the gene-based nutrition is the bright future of nutritional science. Such switch will not only help disease prevention but also will improve the general health. Advancement in the genome-wide studies that associate not only SNPs but also the large genetic variation like insertion-deletion-inversion to the nutrient-related health outcomes are critical for rapid advancement in this field. Improved accuracy in testing approaches such as the DNA base microarray will help in detecting the susceptible individual and will reduced false-positive and false-negative results. Uncovering more underlining mechanisms that clarify how nutrients affect gene expressions will guide the individually tailored nutritional advice.

References

Adjakly M et al (2013) Genistein and daidzein: different molecular effects on prostate cancer. Anticancer Res 33(1):39–44

Afshin A et al (2019) Health effects of dietary risks in 195 countries, 1990–2017: a systematic analysis for the Global Burden of Disease Study 2017. Lancet 393(10184):1958–1972

Ağagündüz D, Gezmen-Karadağ M (2019) Association of FTO common variant (rs9939609) with body fat in Turkish individuals. Lipids Health Dis 18(1):1–12

Alam I et al (2019) Relationship of nutrigenomics and aging: involvement of DNA methylation. J Nutr Intermed Metab 16:100098

Allayee H et al (2008) Nutrigenetic association of the 5-lipoxygenase gene with myocardial infarction. Am J Clin Nutr 88(4):934–940

Ardekani AM, Jabbari S (2009) Nutrigenomics and cancer. Avicenna J Med Biotechnol 1(1):9

Astley SB (2007) An introduction to nutrigenomics developments and trends. In: Genes and nutrition. Springer

Atanasov AG et al (2015) Discovery and resupply of pharmacologically active plant-derived natural products: a review. Biotechnol Adv 33(8):1582–1614

Blander G, Guarente L (2004) The Sir2 family of protein deacetylases. Annu Rev Biochem 73(1): 417–435

Bordoni L, Gabbianelli R (2019) Primers on nutrigenetics and nutri(epi)genomics: origins and development of precision nutrition. Biochimie 160:156–171

Braicu C et al (2017) Nutrigenomics in cancer: revisiting the effects of natural compounds. In: Seminars in cancer biology. Elsevier

Castaner O et al (2012) Protection of LDL from oxidation by olive oil polyphenols is associated with a downregulation of CD40-ligand expression and its downstream products in vivo in humans. Am J Clin Nutr 95(5):1238–1244

Caulfield T et al (2010) Direct-to-consumer genetic testing: good, bad or benign? Clin Genet 77(2): 101–105

Choi EJ, Kim G-H (2013) Antiproliferative activity of daidzein and genistein may be related to ERα/c-erbB-2 expression in human breast cancer cells. Mol Med Rep 7(3):781–784

Chuengsamarn S et al (2012) Curcumin extract for prevention of type 2 diabetes. Diabetes Care 35(11):2121–2127

Control, C.F.D. and Prevention (2009) The power of prevention: chronic disease... the public health challenge of the 21st century. National Center for Chronic Disease Prevention and Health Promotion, Centers for Disease Control and Prevention, Atlanta, GA

Corella D, Ordovas JM (2009) Nutrigenomics in cardiovascular medicine. Circ Cardiovasc Genet 2(6):637–651

De Santis S et al (2019) Extra virgin olive oil: lesson from nutrigenomics. Nutrients 11(9):2085

Del Follo-Martinez A et al (2013) Resveratrol and quercetin in combination have anticancer activity in colon cancer cells and repress oncogenic microRNA-27a. Nutr Cancer 65(3):494–504

Di Renzo L et al (2019) Role of personalized nutrition in chronic-degenerative diseases. Nutrients 11(8):1707

Du Q et al (2013) Synergistic anticancer effects of curcumin and resveratrol in Hepa1-6 hepatocellular carcinoma cells. Oncol Rep 29(5):1851–1858
Duicu C et al (2016) FTO rs 9939609 SNP is associated with adiponectin and leptin levels and the risk of obesity in a cohort of romanian children population. Medicine 95(20)
Edenberg HJ (2007) The genetics of alcohol metabolism: role of alcohol dehydrogenase and aldehyde dehydrogenase variants. Alcohol Res Health 30(1):5
Fawcett KA, Barroso I (2010) The genetics of obesity: FTO leads the way. Trends Genet 26(6): 266–274
Fraga MF et al (2005) Epigenetic differences arise during the lifetime of monozygotic twins. Proc Natl Acad Sci 102(30):10604–10609
Frary CD, Johnson RK, Wang MQ (2005) Food sources and intakes of caffeine in the diets of persons in the United States. J Am Diet Assoc 105(1):110–113
Fu J, Hofker M, Wijmenga C (2015) Apple or pear: size and shape matter. Cell Metab 21(4): 507–508
Gaboon NE (2011) Nutritional genomics and personalized diet. Egypt J Med Hum Genet 12(1)
Garrod A (1902) The incidence of alkaptonuria: a study in chemical individuality. Lancet 160(4137):1616–1620
German JB (2005) Genetic dietetics: nutrigenomics and the future of dietetics practice. J Am Diet Assoc 105(4):530–531
Hardman WE (2014) Diet components can suppress inflammation and reduce cancer risk. Nutr Res Pract 8(3):233
Hardy DS et al (2020) Ancestry specific associations of FTO gene variant and metabolic syndrome: a longitudinal ARIC study. Medicine 99(6)
Hass BS et al (1993) Effects of caloric restriction in animals on cellular function, oncogene expression, and DNA methylation in vitro. Mutat Res 295(4–6):281–289
Hawkinson AK (2007) Nutrigenomics and nutrigenetics in whole food nutritional medicine. Townsend Lett Exam Altern Med 283:102–104
He F-J, Chen J-Q (2013) Consumption of soybean, soy foods, soy isoflavones and breast cancer incidence: differences between Chinese women and women in Western countries and possible mechanisms. Food Sci Human Wellness 2(3–4):146–161
Hoffmann TJ et al (2011) Next generation genome-wide association tool: design and coverage of a high-throughput European-optimized SNP array. Genomics 98(2):79–89
Hua X et al (2016) Association among dietary flavonoids, flavonoid subclasses and ovarian cancer risk: a meta-analysis. PLoS One 11(3):e0151134
Jones PA, Laird PW (1999) Cancer-epigenetics comes of age. Nat Genet 21(2):163–167
Kaput J et al (2007) Application of nutrigenomic concepts to type 2 diabetes mellitus. Nutr Metab Cardiovasc Dis 17(2):89–103
Kaur K, Allahbadia G, Singh M (2018) Impact of nutrigenomics on various metabolic disorders in relation to life style alteration. Austin J Nutri Food Sci 6(1):1100
Kikuno N et al (2008) Genistein mediated histone acetylation and demethylation activate tumor suppressor genes in prostate cancer. J Urol 179(4S):45–45
Kim GT, Lee SH, Kim YM (2013) Quercetin regulates sestrin 2-AMPK-mTOR signaling pathway and induces apoptosis via increased intracellular ROS in HCT116 colon cancer cells. J Cancer Prev 18(3):264
Kohlmeier M (2012) Nutrigenetics: applying the science of personal nutrition. Academic Press
Konstantinidou V et al (2010) In vivo nutrigenomic effects of virgin olive oil polyphenols within the frame of the Mediterranean diet: a randomized controlled trial. FASEB J 24(7):2546–2557
Kühn AB et al (2016) FTO gene variant modulates the neural correlates of visual food perception. NeuroImage 128:21–31
Liu RH (2004) Potential synergy of phytochemicals in cancer prevention: mechanism of action. J Nutr 134(12):3479S–3485S
Liu S et al (2019) Apolipoprotein E gene polymorphism and the risk of cardiovascular disease and type 2 diabetes. BMC Cardiovasc Disord 19(1):1–6

Martins I et al (2006) Apolipoprotein E, cholesterol metabolism, diabetes, and the convergence of risk factors for Alzheimer's disease and cardiovascular disease. Mol Psychiatry 11(8):721–736
Mazo DF et al (2019) Validation of PNPLA3 polymorphisms as risk factor for NAFLD and liver fibrosis in an admixed population. Ann Hepatol 18(3):466–471
Medina-Gomez C et al (2015) Challenges in conducting genome-wide association studies in highly admixed multi-ethnic populations: the Generation R Study. Eur J Epidemiol 30(4):317–330
Meeran SM et al (2011) A novel prodrug of epigallocatechin-3-gallate: differential epigenetic hTERT repression in human breast cancer cells. Cancer Prev Res 4(8):1243–1254
Micha R et al (2017) Association between dietary factors and mortality from heart disease, stroke, and type 2 diabetes in the United States. JAMA 317(9):912–924
Miggiano G, De Sanctis R (2006) Nutritional genomics: toward a personalized diet. Clin Ter 157(4):355–361
Milagro F et al (2013) Dietary factors, epigenetic modifications and obesity outcomes: progresses and perspectives. Mol Asp Med 34(4):782–812
Milenkovic D et al (2012) Modulation of miRNA expression by dietary polyphenols in apoE deficient mice: a new mechanism of the action of polyphenols. PLoS One 7(1):e29837
Müller M, Kersten S (2003) Nutrigenomics: goals and strategies. Nat Rev Genet 4(4):315–322
Mullins VA et al (2020) Genomics in personalized nutrition: can you "eat for your genes"? Nutrients 12(10):3118
Nandakumar V, Vaid M, Katiyar SK (2011) (−)-Epigallocatechin-3-gallate reactivates silenced tumor suppressor genes, Cip1/p21 and p 16 INK4a, by reducing DNA methylation and increasing histones acetylation in human skin cancer cells. Carcinogenesis 32(4):537–544
Nehlig A (2018) Interindividual differences in caffeine metabolism and factors driving caffeine consumption. Pharmacol Rev 70(2):384–411
Nelson MR et al (2004) Large-scale validation of single nucleotide polymorphisms in gene regions. Genome Res 14(8):1664–1668
Omene C et al (2013) Propolis and its active component, caffeic acid phenethyl ester (CAPE), modulate breast cancer therapeutic targets via an epigenetically mediated mechanism of action. J Cancer Sci Ther 5(10):334
Ozturk G et al (2012) The anticancer mechanism of caffeic acid phenethyl ester (CAPE): review of melanomas, lung and prostate cancers. Eur Rev Med Pharmacol Sci 16(15):2064–2068
Paredes-Gonzalez X et al (2014) Apigenin reactivates Nrf2 anti-oxidative stress signaling in mouse skin epidermal JB6 P+ cells through epigenetics modifications. AAPS J 16(4):727–735
Peregrin T (2001) The new frontier of nutrition science: nutrigenomics. J Acad Nutr Diet 101(11): 1306
Petric RC et al (2015) Phytochemicals modulate carcinogenic signaling pathways in breast and hormone-related cancers. Onco Targets Ther 8:2053
Piroddi M et al (2017) Nutrigenomics of extra-virgin olive oil: a review. Biofactors 43(1):17–41
Qian F et al (2019) Association between plant-based dietary patterns and risk of type 2 diabetes: a systematic review and meta-analysis. JAMA Intern Med 179(10):1335–1344
Reen JK, Yadav AK, Singh J (2015) Nutrigenomics: concept, advances and applications. Asian J Dairy Food Res 34(3):205–212
Remely M et al (2015a) Nutriepigenomics: the role of nutrition in epigenetic control of human diseases. Curr Opin Clin Nutr Metab Care 18(4):328–333
Remely M et al (2015b) Therapeutic perspectives of epigenetically active nutrients. Br J Pharmacol 172(11):2756–2768
Sachse C et al (2003) Polymorphisms in the cytochrome P450 CYP1A2 gene (CYP1A2) in colorectal cancer patients and controls: allele frequencies, linkage disequilibrium and influence on caffeine metabolism. Br J Clin Pharmacol 55(1):68–76
Salzberg SL (2018) Open questions: how many genes do we have? BMC Biol 16(1):1–3
Satija A, Hu FB (2018) Plant-based diets and cardiovascular health. Trends Cardiovasc Med 28(7): 437–441

Shu L et al (2011) Epigenetic CpG demethylation of the promoter and reactivation of the expression of Neurog1 by curcumin in prostate LNCaP cells. AAPS J 13(4):606–614
Shungin D et al (2015) New genetic loci link adipose and insulin biology to body fat distribution. Nature 518(7538):187–196
Singh AN, Baruah MM, Sharma N (2017) Structure based docking studies towards exploring potential anti-androgen activity of selected phytochemicals against prostate cancer. Sci Rep 7(1):1–8
Stover PJ, Garza C (2002) Bringing individuality to public health recommendations. J Nutr 132(8): 2476S–2480S
Tolba MF et al (2013) Caffeic acid phenethyl ester synergistically enhances docetaxel and paclitaxel cytotoxicity in prostate cancer cells. IUBMB Life 65(8):716–729
Upadhyay S, Dixit M (2015) Role of polyphenols and other phytochemicals on molecular signaling. Oxidative Med Cell Longev 2015
Vardi A et al (2010) Soy phytoestrogens modify DNA methylation of GSTP1, RASSF1A, EPH2 and BRCA1 promoter in prostate cancer cells. In Vivo 24(4):393–400
Waltenberger B et al (2016) Natural products to counteract the epidemic of cardiovascular and metabolic disorders. Molecules 21(6):807
Weng C-J, Yen G-C (2012) Chemopreventive effects of dietary phytochemicals against cancer invasion and metastasis: phenolic acids, monophenol, polyphenol, and their derivatives. Cancer Treat Rev 38(1):76–87
Wubetu GY et al (2016) Epigallocatechin gallate hinders human hepatoma and colon cancer sphere formation. J Gastroenterol Hepatol 31(1):256–264
Yang Y et al (2007) Effect of apolipoprotein E genotype and saturated fat intake on plasma lipids and myocardial infarction in the Central Valley of Costa Rica. Hum Biol 79(6):637–647
Yates Z, Lucock M (2003) Interaction between common folate polymorphisms and B-vitamin nutritional status modulates homocysteine and risk for a thrombotic event. Mol Genet Metab 79(3):201–213
Yi SV, Goodisman MA (2021) The impact of epigenetic information on genome evolution. Philos Trans R Soc B 376(1826):20200114
Yoo J-H, Park S-C (2000) Low plasma folate in combination with the 677 C→ T methylenetetrahydrofolate reductase polymorphism is associated with increased risk of coronary artery disease in Koreans. Thromb Res 97(2):77–84
Zhang N (2015) Epigenetic modulation of DNA methylation by nutrition and its mechanisms in animals. Anim Nutr 1(3):144–151

Chapter 4
Nutrigenomics of Aging

Nelma M. Palminha, Sara Ahmed Hashish, Eman Rabie, and Sherine Khalafalla Saber

4.1 Introduction

Aging is the progressive tissue dysfunction ultimately causing deterioration of organs/systems (Niccoli and Partridge 2012). Many diseases have been linked to aging, for example, cardiovascular disorders, neurodegenerative diseases, cancer, type 2 diabetes, age-related cachexia, glaucoma, renal dysfunction, osteoporosis, and osteoarthritis. However, aging research is still yet to understand the aging process. The concept of aging hallmarks serves to identify commonly occurring age-related cellular and molecular mechanisms.

Aging hallmarks can be categorized into three groups: (1) primary causes of age-associated damage including changes in chromatin structure, epigenetic alterations, and impaired protein homeostasis; (2) antagonistic responses to the damage which include decreased nutrient sensing, mitochondrial dysfunction, and cellular senescence; and lastly (3) integrative consequences of the responses which include stem cell exhaustion and altered intracellular communication (López-Otín et al. 2013).

The aim of this chapter is to shed light on the link between nutrigenomics and aging. Nutrigenomics is an emerging field that studies how diet impacts the genome.

N. M. Palminha (✉)
Department of Molecular Biology and Biotechnology, The Healthy Lifespan Institute, Sheffield, UK

The Institute of Neuroscience, University of Sheffield, Sheffield, UK
e-mail: nmpereirapalminha1@sheffield.ac.uk

S. A. Hashish
The MPH Program, Institute of Global Health and Human Ecology, The American University in Cairo, New Cairo, Egypt

E. Rabie · S. K. Saber
The Biotechnology Program, The American University in Cairo, New Cairo, Egypt

M. Salama (ed.), *Nutrigenomics and the Brain*, Nutritional Neurosciences,
https://doi.org/10.1007/978-981-16-9205-5_4

Therefore, we first clarify the mechanisms of cellular senescence, explain how telomeres are involved in aging, and explore how the epigenome changes during the aging process. In addition, we will address how these mechanisms connect in driving genomic instability during aging. Finally, we will delve into how the aging process affects physiology and explain how nutrition can influence several aspects associated with aging.

4.2 Cellular Senescence

Cellular senescence was first observed among fibroblast cells of primary cultures when over time their division slows down and eventually comes to a permanent stoppage (Hayflick 1965). Cellular senescence is a cellular phenomenon that induces a stable and usually irreversible growth arrest accompanied by distinct phenotypic alterations such as remodeling of the chromatin, metabolic reprogramming, increase in autophagy, and the production of a proinflammatory senescence secretome, also known as senescence-associated secretory phenotype or SASP (Kuilman et al. 2010; Salama et al. 2014). It is important to know that cellular senescence is a response to various types of stressors and/or damage according to which it can be subtyped, for example, "replicative senescence" which is induced by telomeric erosion after multiple replications accompanied by reduction in the cell's proliferative potential. Another subtype is "premature senescence" which can be stress-induced by DNA damage, oxidative stress, and/or mitochondrial injury or therapy-induced, for example, by chemotherapeutics or radiation. The term "oncogene-induced senescence" refers to the induction of senescence by anomalous oncogene activation which results in hyperproliferation and subsequent proliferation arrest (Song et al. 2020). Physiologically, "programmed senescence" plays a pro-regenerative role in synchronization with apoptosis during normal development, particularly in embryogenesis (Muñoz-Espín et al. 2013; Storer et al. 2013). Also, senescence is indispensable for tissue homeostasis in adult tissue to facilitate cell turnover and tissue remodeling (Song et al. 2020). The physiological role of senescence also involves its strong tumor suppression capabilities, limiting cancer initiation (Hanahan and Weinberg 2011; Storer et al. 2013). However, accumulation of senescent cells has disadvantageous effects through the proinflammatory SASP which provides a suitable environment for enhancing proliferative potential of cancerous cells (Krtolica et al. 2001; Coppé et al. 2010). It is also reported that senescent cells contribute to epithelial to mesenchymal transition (EMT) (Coppé et al. 2008). Furthermore, elimination of senescent cells was found to delay tumor onset and decrease metastasis (Baker et al. 2016; Demaria et al. 2017). This double-edged role of cellular senescence is what makes the therapeutic targeting of key senescent pathways difficult and poses potential risks of side effects.

4.2.1 Senescence-Mediated Arrest

The hallmark of cellular senescence is the growth arrest that prevents the proliferation of aged, damaged, or transformed cells. Two key tumor suppression pathways, p16INK4a/Rb and p53/p21CIP1, are activated in senescence-mediated growth arrest in response to DNA damage whether DNA damage was induced by oxidative stress, telomere attrition, oncogene overexpression, or metabolic dysfunction (McHugh and Gil 2018).

Both pathways eventually inhibit CDK4/6 which in turn hypophosphorylates Rb with subsequent exit from the cell cycle. The activation of chromatin changes; particularly elevated γH2Ax and 53BP1 deposition in chromatin activates the kinases cascade of ATM and ATR and then checkpoints CHK1 and CHK2, ultimately activating p53. p53 in turn inhibits p21 via increasing cyclin-dependent kinase inhibitor (p21CIP1) gene transcription, eventually blocking CDK4/6 to hypophosphorylate Rb and enforcing cell cycle exit (d'Adda di Fagagna 2008; Fumagalli et al. 2012). In contrast to transient p53 induction that leads to quiescence and repair, persistent induction of p53/p21CIP1 with subsequent permanent arrest of cell cycle occurs during senescence. The reason for this persistence is that DNA damage occurs in the repair-resistant DNA segments with chromatin alterations reinforcing senescence (DNA-SCARS), e.g., telomeres (Rodier et al. 2011; Fumagalli et al. 2012; Salama et al. 2014; Kruiswijk et al. 2015). p53 induction is also controlled by other upstream regulators, such as forkhead box protein O4 (FOXO4) which activates nuclear localization and transcription of p53; indeed FOXO4 inhibitor delayed aging phenotype in mice (Baar et al. 2017).

The activation of p16INK4a/Rb pathway involves the induction of INK2/ARF locus, a 35-kb locus comprising of two tumor suppressor cyclin-dependent kinase inhibitors genes: CDKN2A which encodes p16INK4a and ARF and CDKN2B which encodes p15INK4b. Both p16INK4a and p15INK4b can inhibit CDK4/6, enforcing cell cycle exit. ARF serves in crosstalk with p53/p21CIP1 pathway; it is an inhibitor of the ubiquitin ligase MDM2 (an upstream inhibitor of p53) and thus, contributes to p53 activation. The expression ARF itself is regulated by p53 in a negative feedback loop (Harris and Levine 2005). Normally, the INK4/ARF locus silenced by H3K27 methylation controlled by polycomb repressive complexes (PRCs) (Bracken et al. 2007). In senescence, the H3K27 histone demethylase (JMJD3) activates the INK4/ARF locus (Krishnamurthy et al. 2004; Agger et al. 2009; Barradas et al. 2009; Burd et al. 2013). Increasing gene dosage of INK4/ARF locus had antiaging effect on mice and extended their longevity. Delayed aging was suggestively attributed to preferring quiescence and slowing down proliferation (Matheu et al. 2009). The INK2/ARF locus is commonly mutated in cancer, and many of its DNA sequence variants were associated with atherosclerosis, diabetes, and other age-related pathologies (Gil and Peters 2006; Kim and Sharpless 2006; Jeck et al. 2012).

4.2.2 Senescence-Associated Secretory Phenotype (SASP)

In addition to growth arrest, another major hallmark for senescent phenotype is the SASP, also termed senescence-messaging secretome which comprises a plethora of soluble factors including proapoptotic factors and proinflammatory cytokines, chemokines, and growth factors (Kuilman and Peeper 2009; Coppé et al. 2010). IL-1 signaling is an important SASP component; IL-1α expression can reproduce SASP in vitro resulting in senescence (Acosta et al. 2013). The secretome is cell type dependent and differs according to the type of senescent inducer; however, common mechanisms and SASP regulators do exist. For example, SASP was found to be regulated by the transcription factors nuclear factor κB (NF-kB) and CCAAT/ enhancer-binding protein beta (c/EBPβ) and GATA binding protein 4 (GATA4) (Acosta et al. 2008; Kuilman et al. 2008; Kang et al. 2015). Moreover, signaling pathways, especially those including cytoplasmic kinases regulate SASP, e.g., p38α MAPK, mTOR, and TAK1. Rapamycin a strong SASP inhibitor is a known mTOR inhibitor; mTOR regulates the translation of IL-1α and degrades transcripts of various SASP components via activation of the RNA binding protein ZFP36L1 (Herranz et al. 2015; Laberge et al. 2015; Zhang et al. 2018). Recently, upstream SASP inducers were investigated; the presence of cytoplasmic chromatin fragments activates cGAS/GAMP-STING pathway, downregulates cytoplasmic DNases, and induces IL-1 inflammatory response (Coppé et al. 2010; Dou et al. 2017; Glück et al. 2017; Yang et al. 2017; Takahashi et al. 2018). Senescence is accompanied by global remodeling of the genome's enhancer landscape, particularly, the activation of super enhancers (SEs) proximal to genes encoding key SASP components (Tasdemir et al. 2016). SASP components function in the activation and recruitment of immune cells, e.g., macrophages, natural killer cells to eliminate senescent cells which has a tumor suppressive role especially in tumor initiation stages (Xue et al. 2007; Kang et al. 2011). Inhibition of the bromodomain protein (BRD4) a chromatin reader that binds to the aforementioned SEs represses this tumor suppression role (Tasdemir et al. 2016). Conversely, other SASP components can activate tumorigenesis, e.g., the role of VEGF and CCL5 in angiogenesis and the role of GROα and osteopontin in tumor growth (Krtolica et al. 2001; Coppé et al. 2006; Eyman et al. 2009; Pazolli et al. 2009). Furthermore, immature immunosuppressive myeloid cells are recruited by SASP in liver and prostate cancers (Di Mitri et al. 2014; Eggert et al. 2016). Also, senescent tumor cells act via SASP to promote invasion and metastasis in thyroid cancer (Kim et al. 2017).

The "senescence-mediated inflammasome" in senescent cells can spread senescence in autocrine and paracrine fashion. TGF-β family ligands are components of SASP that were found to be drivers of paracrine signaling as well as regulators of p15INK4b and p21CIP1, the key players of growth arrest (Acosta et al. 2013). A number of SASP components, e.g., IL-8, GROα, IL-6, and IGBP-7, themselves exacerbate the senescent phenotype (Kuilman et al. 2008; Wajapeyee et al. 2008; Acosta et al. 2013).

4.2.3 Biomarkers of Cellular Senescence

Cellular senescence plays a role in physiological and pathological contexts; consequently finding a single robust and consistent biomarker for senescence is problematic (Muñoz-Espín and Serrano 2014; Childs et al. 2015). Adding to the complexity of studying senescence is the different key players according to the senescent inducer and the type of the affected cell; thus a multi-marker approach might be considered (Hernandez-Segura et al. 2017). The majority of studies use senescence-associated β-galactosidase (SA-β-Gal) staining or the absence of proliferation markers such as Ki67. Lipofuscin is also used due to its cytoplasmic accumulation in senescent cells (Sharpless and Sherr 2015). p16INK4a is regarded an aging biomarker; its expression increased with aging across different cell types including the islet of Langerhans, renal cortex, and adipose tissue (Krishnamurthy et al. 2004; Melk et al. 2004; Lomas et al. 2012; Baker et al. 2016; Helman et al. 2016). Transplantation of p16INK4a−/−hematopoietic stem cells (HSCs) in mice reduced age-associated HSC apoptosis under stress and enhanced repopulation of injured tissue (Janzen et al. 2006).

4.2.4 Effect of Accumulation of Senescent Cells

Elimination of senescent cells reduces inflammatory markers (IL-6 and IL-1β). This effect has been seen in an aging kidney, spleen, lung, liver, and heart and arthritic knee (Baker et al. 2016; Baar et al. 2017; Jeon et al. 2017). Inflammatory SASP components increase in aging tissues, and their inhibition has been shown to cause resistance to progeroid phenotypes (Tilstra et al. 2012). Chronic inflammation induced by SASP or "inflammaging" has damaging role seen in many clinical studies (Libby et al. 2002; Brunt et al. 2009; Dinarello et al. 2010; Franceschi and Campisi 2014; Balestro et al. 2016). Increase in SASP components in senescent cells and increase in the number of senescent cells facilitate their elimination via immune system (Kang et al. 2011). However, immune system functions decline through aging process which could contribute to lessened senescent cells' clearance and sustain inflammation. Some studies have linked the elimination of senescent cells to reduced inflammatory processes (Baker et al. 2016; Jeon et al. 2017).

4.2.5 Metabolic Dysfunction and Cellular Senescence

Metabolic dysfunction is a damage process that causes aging and drives senescence (López-Otín et al. 2013). Metabolic pathways, e.g., mTOR (mechanistic target of rapamycin), or insulin pathway receives input of growth and nutrient signals to control cellular functions, e.g., synthesis of proteins and lipids, autophagy, and metabolism. Calorie restriction inhibits mTOR signaling, deters aging decline, and is linked to increased health as well as life span (Mitchell et al. 2016; Saxton and Sabatini 2017). mTOR signaling regulates SASP, senescence-mediated arrest, and autophagy (Herranz et al. 2015; Laberge et al. 2015). Another example of the link between metabolic dysfunction and senescence is sirtuins. Sirtuins, such as SIRT1 and SIRT6, are ribosyltransferases that regulate metabolism and DNA repair. SIRT1 causes p53 degradation via its deacetylation, while SIRT6 deacetylates H3K18 averting mitotic errors and deterring senescence (Solomon et al. 2006; Houtkooper et al. 2012; Tasselli et al. 2016).

4.3 Telomeres and Aging

4.3.1 Structure and Function of Telomeres

Telomeres are highly conserved nucleotide repeats that cap both ends of each chromosome to maintain genomic integrity (Srinivas et al. 2020). Telomerase holoenzyme is a unique nucleoprotein structure where an array of telomere-related proteins binds to telomeric DNA to construct special protein/DNA complexes to ensure genomic stability. This telomerase complex that is composed of telomeric reverse transcriptase (TERT), telomeric RNA component (TERC), and other factors adds telomeric repeats to the ends of chromosomes (Chan et al. 2010). However, during normal cellular processes, such as cell division, telomeric DNA loses a small part, forcing the cell to undergo senescence and/or apoptosis when telomere length hits a certain point (Shammas 2011). Telomeres length may thus serve as a biological marker to determine the organism's life span.

4.3.2 Telomeres, Lifestyle, and Longevity

Telomeres decrease in their length as one ages (Valdes et al. 2005). In humans, telomere length declines at a rate of 24.8–27.7 base pairs per year (Brouiette et al. 2003). A shorter telomere length than the average length in a certain age group has been correlated with the incidence of age-related diseases and/or decreased life span in individuals (Cawthon et al. 2003). There are several factors that may contribute to telomere shortening, thus leading to aging. Some of these factors are genetic, epigenetic and environmental, body weight, social and economic status, and exercise, along with smoking (Shammas 2011). Shammas further highlighted that there are specific lifestyle practices such as lack of exercise, unhealthy food habits, smoking, and obesity that can trigger telomere shortening, inducing premature death or age-associated health issues, including coronary heart disease, diabetes, heart failure, and osteoporosis.

4.3.3 ROS on Telomeres

The impact of aging on DNA damage in human cells has been investigated in several studies (Bautista-Niño et al. 2016). Genome damage piles up over time due to lower DNA repair capacity, lower chromosome segregation, and cell cycle checkpoint efficacy (Moskalev et al. 2013). Endogenous gentoxins and inadequate nutrition, along with other factors and lifestyles, may also contribute to an increase in baseline genome damage and accelerated telomere shortening (Vidacek et al. 2017). Most of these factors indirectly impact aging through free radical formation and oxidative stress. This suggests that since reactive oxygen species (ROS) predominantly induce age-related damage at the cellular and tissue levels, a decline in calorie intake may lead to a reduction in energy metabolism and ROS generation, resulting in a diminished cell and tissue damage. Indeed, ROS strongly influences replicative senescence and aging via accelerated telomere shortening caused primarily due to accumulation of nicks in the G-rich strand (Vidacek et al. 2017).

Under physiological conditions, the mitochondria are the main source of ROS, and it has been reported that senescence of human cells is associated with disabled mitochondria and shorter telomeres (Passos et al. 2007). Excessive production of ROS results in oxidative stress which can lead to oxidation of biomolecules including lipids, proteins, and DNA over time. This loss of DNA structure cannot be reversed (Lieber and Karanjawala 2004). Evidently, there is a valid reason why chronic oxidative stress is considered one of the leading causes of various metabolic and neurodegenerative diseases and chronic inflammation. Inflammation causes accelerated white blood cell turnover and additional telomere attrition resulting in accelerated aging (Vidacek et al. 2017).

4.4 Epigenetic Changes During Aging

The DNA double-helix is organized into high-order complex structures—the chromatin, a nucleoprotein complex that allows the packaging of the DNA inside the nucleus (Benayoun et al. 2015). The basic unit of the chromatin is the nucleosome, which consists in 145–147 bp of DNA wrapped around a histone octamer, consisting of dimers of the core histones H2A, H2B, H3, and H4 (Zhou et al. 2019). By packing the DNA into these structures, the chromatin modulates gene expression by managing the access to DNA to transcription factors (Li et al. 2018; Zhou et al. 2021). The chromatin exists in two states: the heterochromatin is a condensed and transcriptionally inactive state, and the euchromatin is a decondensed chromatin stage and therefore with a conformation accessible to transcription factors (Benayoun et al. 2015; Machida et al. 2018). The dynamics of the chromatin structure is highly regulated by enzymes that catalyze histone posttranslation modifications (Baldensperger et al. 2020).

During the aging process, the chromatin suffers alterations (Kane and Sinclair 2019). These modifications in the chromatin homeostasis affect many cellular processes that affect organismal life span, including increased genomic instability, altered DNA replication, and changes in gene expression (Pal and Tyler 2016). The latter is due to loss of transcriptional control that ultimately led to the deterioration observed in aged organisms (Stegeman and Weake 2017). These epigenetic alterations encompass one of the hallmarks of aging and can be influenced by both internal factors and external stimuli, including nutrition (Benayoun et al. 2015; Pal and Tyler 2016). Since these changes do not alter the DNA sequence, they can explain some of the variation in adult life span (O'Sullivan and Karlseder 2012; Yi and Kim 2020).

In aged cells, epigenetic regulatory changes include heterochromatin loss and reorganization, accompanied by reduction of core histones, altered DNA methylation patterns, and global changes in histone posttranslational modifications (Pal and Tyler 2016).

Increased transcriptional activation of otherwise silenced genes is a characteristic of aging cells and the main consequence of age-related changes in the epigenome (Kane and Sinclair 2019). Repressive chromatin depends on the presence of heterochromatin protein (HP1), the linker histone H1, and the trimethylation of the lysine 9 on histone H3 (H3K9me3) (Machida et al. 2018; Pérez et al. 2018; Nicetto et al. 2019). Together, these factors stabilize the heterochromatin complex and play a pivotal role in silencing genomic regions (Maison and Almouzni 2004; Tsurumi and Li 2012; Nicetto and Zaret 2019; Lee et al. 2020). The age-related global loss of transcriptional silencing is caused by alterations in the chromatin state that results in a noncontrolled access to the genetic material (Larson et al. 2012). "The heterochromatin loss model of aging," proposed in 1997 by Villeponteau, was one of the earliest models to associate epigenetic changes as a marker of aging. According to this model, the heterochromatin is disrupted during aging, leading to changes in nuclear architecture and suppression of transcription repression, thus causing aberrant expression patterns (Villeponteau 1997).

Supporting this model, several studies performed in *C. elegans*, *D. melanogaster*, and senescent human fibroblasts have described loss of the heterochromatin markers, H3K9me3, and reduced levels of HP1 (Haithcock et al. 2005; Brandt et al. 2008; Larson et al. 2012; Ni et al. 2012; O'Sullivan and Karlseder 2012; Ivanov et al. 2013). Other signs of heterochromatin loss observed in cellular models include increased nuclear area, accompanied by DNA decondensation (Lee et al. 2020).

Models of premature aging diseases also report heterochromatin loss (Yi and Kim 2020). Like models of chronological aging, which also exhibit defects in lamin A, cells from Hutchinson-Gilford progeria (HGPS) patients exhibit enlarged nuclei, which indicate heterochromatin loss. These cells also show loss of H3K9me3 and reduced HP1 expression (Scaffidi and Misteli 2006; Shumaker et al. 2006).

Werner syndrome (WS) models have also demonstrated loss of heterochromatin. Both WS mice and human stem cell models showed decreased of SUV39H1, a H3K9me3 methyltransferase. As described in other aging models, *WRN* knockout stem cells also exhibit atypical nuclear envelope and enlarged nucleus. Additionally,

HP1α overexpression rescued premature senescence and promoted heterochromatin stabilization in *WRN* knockout cells, further showing the role of loss of heterochromatin in aging (Zhang et al. 2015).

Epigenetic changes are also believed to contribute to replicative senescence (Booth and Brunet 2016; Kane and Sinclair 2019). During cellular senescence, the heterochromatin is redistributed, forming senescence-associated heterochromatin foci (SAHF) and consequent silencing of genes mostly associated with proliferation (Narita et al. 2006; Sadaie et al. 2013). A global loss of heterochromatin is also observed in senescent cells, associated with loss of gene silencing and expression of inactivated genes (Lee et al. 2020).

In addition to heterochromatin decay and remodeling, reduced expression of core histones and replacement of canonical histones with histone variants are observed during aging (Pal and Tyler 2016; Kane and Sinclair 2019; Yi and Kim 2020). Studies in budding yeast revealed a decrease of the core histones during replicative aging. This reduction in histones was accompanied by loss of transcriptional repression (Hu et al. 2014). Conversely, extended replicative life span and transcription silencing was achieved after supplying yeast with extra histone proteins, either by inactivation of the histone information regulator (Hir) complex or by ectopic expression of histones H3 and H4 through an inducible promoter system (Feser et al. 2010).

Reduced histone levels and chromatin reorganization were also observed during replicating aging in human diploid fibroblasts (O'Sullivan and Karlseder 2012). This reduction was due to the chronic DNA damage associated with shortening of telomeres, causing histone biosynthesis downregulation (O'Sullivan and Karlseder 2012; Pal and Tyler 2016).

During yeast aging, several indicators of genomic alterations were also noticed, such as the presence of DNA breaks, chromosomal translocations, insertions of mitochondrial DNA into nuclear DNA, and retrotransposition (Hu et al. 2014; Kane and Sinclair 2019). Similarly, aged cells from several aging models, including cells from progeria models and senescent human cells, show persistent levels of DNA and the presence of retrotransposable elements (Maxwell et al. 2011).

Another form of transcriptional regulation is achieved through DNA methylation (Pal and Tyler 2016). DNA methylation occurs mainly at CpG sites or repetitive regions of the genome, such as LINE-1 sequences. CpG methylation at promoters is associated with repressive chromatin structure (Kane and Sinclair 2019). Age-related changes in DNA methylation patterns have been reported (Li et al. 2017; Kane and Sinclair 2019). While a general loss of DNA methylation occurs during aging due to an age-dependent decrease in the levels of the DNA methyltransferase DNMT1, promoters of highly expressed genes, which lack methylation (CpG islands), become hypermethylated (Jung and Pfeifer 2015). This leads to age-related changes in gene expression, which can promote the development of diseases associated with aging (Fenech et al. 2011; Jung and Pfeifer 2015). Decreased DNA methylation of retrotransposable elements is associated with increased risk of cancer development (Pal and Tyler 2016). Retrotransposons are mobile DNA elements that are silenced in young cells. As organisms age, retrotransposable elements are activated due to diminished DNA methylation and consequent loss of transcription repression

(Maxwell 2016). The loss of transcription repression associated with age leads to the transcription of these elements, followed by the reversed transcription of the newly transcribed RNA and integration of the resulting cDNA elsewhere into the genome (Maxwell et al. 2011). Retrotransposons are another form of genomic instability and have increased mutagenic potential, promoting cancer development. Interestingly, increased retrotransposition can also promote neurodegeneration, both a risk factor of aging (Hou et al. 2019).

Histone posttranslation modifications influence the way the DNA is regulated (Lawrence et al. 2016). Histone modifications can affect chromatin dynamics and promote or inhibit the recruitment of a wide variety of factors (Yi and Kim 2020). This way cells regulate several biological processes, such as transcription, replication, or DNA repair. Enzymes that catalyze the addition of removal of histone posttranslation modifications function like a well-oiled machine (O'Sullivan and Karlseder 2012; Pal and Tyler 2016; Kane and Sinclair 2019; Yi and Kim 2020). However, disruptions in the way these enzymes work occur in an age-dependent manner. These alterations differ between organisms, tissues, or even between cells but often result in a more permissive chromatin, leading to aberrant gene expression (Benayoun et al. 2015; Kane and Sinclair 2019).

Histone acetylation negatively influences gene silencing and consequently has a negative impact on longevity (Pal and Tyler 2016). Several studies have shown that in both yeast and mammalian cell models, histone acetylation changes globally with age, leading to the loss of chromatin integrity (Peleg et al. 2016). Multiple reports have demonstrated the importance of the balance between histone acetylation and deacetylation in prolonging life span (Benayoun et al. 2015; Peleg et al. 2016; Kane and Sinclair 2019). For instance, several studies have shown that inhibition of histone deacetylase (HDAC), such as sirtuins, shortens life span (Pal and Tyler 2016). One example is the yeast Sir2, an NAD^+-dependent class III HDAC, which catalyzes the deacetylation of H4K16Ac. Sir2 is implicated in rDNA gene silencing and genomic and telomeric stabilization (Dang et al. 2009). During replicative aging, Sir2 expression decreases, which leads to increased H4K16Ac and consequent genomic instability (Dang et al. 2009). Additionally, Sir2 loss is also associated with sterility, a characteristic of aging organisms. Furthermore, studies have shown that promotion of H4K16Ac increases life span, either by Sir2 upregulation through overexpression of *SIR2* gene or by deletion of *SAS2*, which encodes for the H4K16Ac-specific histone acetyltransferase (HAT) (Kaeberlein et al. 1999; Dang et al. 2009).

In contrast, H3K56Ac, which promotes genomic stability, gene expression, and DNA replication, decreases during yeast aging (Williams et al. 2008; Sen et al. 2016; Kane and Sinclair 2019). However, studies have shown that cells need an optimal amount of H3K56Ac to promote life span extension since both an excess and shortage of this modification shortens life span (Dang et al. 2009; Feser et al. 2010). Aged human fibroblasts also display an overall increase in H4K16Ac and a decrease in H3K56Ac (Feser et al. 2010).

Sirtuins have other substrates apart from histones. For example, during oxidative stress, the mammalian Sir2 ortholog, SIRT1, relocalizes to the sites of DNA breaks

(Alves-Fernandes and Jasiulionis 2019). SIRT1 deacetylates H1K26Ac, a marker of facultative heterochromatin, H3K9Ac and K4K16Ac (Alves-Fernandes and Jasiulionis 2019). It is also associated with the repression of repetitive elements. Due to increased oxidative DNA damage, characteristic of aging cells, SIRT1 redistributes, leading to changes in these histone acetylation markers and consequent chromatin disruption (Kane and Sinclair 2019).

SIRT6 is another Sir2 ortholog that targets H3K9Ac and H3K56Ac. Mice deficient in SIRT6 show premature aging phenotypes (Tasselli et al. 2017). SIRT6 is not only a HDAC but is also responsible for the mono-ADP-ribosylation of KAP1, promoting its interaction with HP1 (Van Meter et al. 2014). This leads to a more repressive heterochromatin and the packaging of L1 LINE retrotransposable elements (Van Meter et al. 2014). The redistribution of SIRT6 during aging to sites of DNA damage leads to changes in the chromatin dynamics with consequent aberrant gene expression patterns, as well as activation of retrotransposons (Van Meter et al. 2014; Xu et al. 2015).

Changes in histone methylation also occur during age. For example, H3K27me3 is associated with a repressive histone state and is altered in aging organisms (Kane and Sinclair 2019). Increasing levels of H3K27me3 were observed in mouse muscle stem cells and brain tissues from SAMP8, a senescent-accelerated mouse models, as well as senescent human lung fibroblasts (Wang et al. 2010; Baumgart et al. 2014). However, decreasing levels of H3K27me3 were described in fibroblasts from HGPS and WS, as well as in *C. elegans* (Shumaker et al. 2006; Ni et al. 2012; Zhang et al. 2015). This might indicate tissue or organismal specificity on the effects this marker has on life span, although its studies point that an overall increase of H3K27me3 is beneficial for life span (Tan et al. 2017).

H3K4me3 levels also change in different aging organisms. In aged *Drosophila*, H3K4me3 decreases, while in mouse hematopoietic stem cells, there is an overall increase. Studies in senescent human fibroblasts indicate locus specificity of this marker, as H3K4me3 increases in some locus while decreases in others.

A global loss of H3K9me3, a hallmark of heterochromatin, is also observed during aging (Nicetto et al. 2019). In contrast, another heterochromatin mark, H4K20me3, shows tissue-specific changes, as its overall levels increase in fibroblasts from HGPS patients, but a decrease of this marker was shown in aging cardiomyocytes (Arancio et al. 2014; Lyu et al. 2018).

4.5 Genome Instability and the Aging Process

Cells suffer tens of thousands of internal and external insults that threat the stability of the genome daily (Abugable et al. 2019). Cells have developed several DNA damage response (DDR) pathways to be able to cope with the lesions inflicted upon the DNA. As cells age, they accumulate DNA damage, and their ability to cope with the increasing repair demand decreases. In fact, evidence suggests a decline of DDR enzymes in aged tissues, including in the hematopoietic system (Gorbunova and

Seluanov 2016). This causes accumulation of unrepaired DNA damage, which increases genomic instability, a known causal factor of aging (Niedernhofer et al. 2018; Abugable et al. 2019).

The link between aging and DNA damage has been established by the existence of rare heritable premature aging syndromes (progeria) caused by defects in DNA repair mechanisms (Liao et al. 2018; Niedernhofer et al. 2018; Abugable et al. 2019). For example, WS is an accelerating aging disorder caused by mutations in *WRN*. *WRN* encodes for a RecQ DNA helicase which is involved in DNA repair and plays a role in telomere maintenance and in protecting cells against replication stress (Oshima et al. 2017). Cockayne syndrome (CS) is caused by mutations in genes that encode for proteins required for transcription-coupled nucleotide excision repair (NER) (Cleaver et al. 1999; Niedernhofer et al. 2018). Another example of a progeroid disorder is Ruijs-Aalfs syndrome (RJALS). RJALS is an autosomal recessive disorder that is caused by mutations in *SPRTN* which encodes for a metalloprotease important for protein-linked DNA breaks (Maskey et al. 2014; Vaz et al. 2016). Another premature aging syndrome, HGPS, is an autosomal-dominant disorder caused by mutations in the *LMNA* gene that results in a truncated form of the lamin A protein (Ahmed et al. 2018). Although lamin A is not directly involved in DNA repair, cells from HGPS patients show increased levels of double-strand breaks (DSBs) due to defects in DDR signaling (Musich and Zou 2011). Lamins are important for maintaining nuclear structure and for chromatin organization, which are crucial for the fine-tune orchestration of DDR and repair signaling (Musich and Zou 2011).

The age-related processes described in the previous sections contribute to the accumulation of DNA damage in an age-dependent manner. In turn, increased DNA damage associated with aging can also influence the abovementioned cellular processes. This indicate an interplay between the different cellular mechanisms that influence and are influenced by the aging process (Fig. 4.1) (Moskalev et al. 2013; Maynard et al. 2015).

An increase of DNA adducts is a frequent finding in aging organisms. One of the most common DNA adducts that accumulates over time is 8-oxoguanine (8-oxoG). 8-oxoG is induced by increased reactive oxygen species (ROS), a by-product of mitochondrial respiration that causes oxidative stress. In fact, the oxidative stress theory of aging proposes that accumulation of oxidative damage is an etiology of aging (Katyal et al. 2007; Madabhushi et al. 2014; Niedernhofer et al. 2018). Other types of damage that increase with age include DNA crosslinks and DSBs breaks. The increase of genomic instability with age contributes to the development of age-related diseases, including cancer, neurodegeneration, atherosclerosis, and cataracts, among others (El-Khamisy and Caldecott 2006; El-Khamisy 2011; Walker et al. 2017; Abugable et al. 2019; Da Silva and Schumacher 2019).

Logically, persistent DNA damage leads to constant activation of the DDR mechanisms. Over time, this alters the cell cycle dynamics, due to the constant need for arresting the cell cycle to promote DNA repair (Crane et al. 2019). When the cells are not able to handle the lack of cell cycle regulation, they can ultimately stop replicating, thus entering in cellular senescence. Alternatively, cells can become

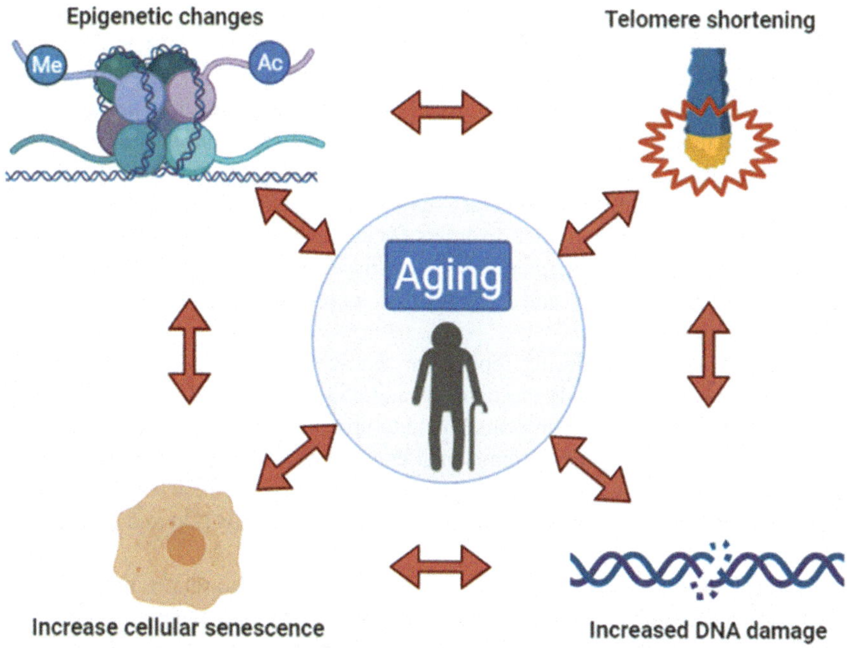

Fig. 4.1 The interplay between different cellular mechanisms during aging

mutagenic, giving rise to cancer. In the case of brain tissue, neurons can die prematurely, leading to neurodegeneration (Petr et al. 2020). In addition, not only increased age-induced DNA damage can lead to increased chromatin relaxation, but the disrupted chromatin dynamics also caused by loss of nucleosomes can itself trigger more genomic instability during aging (Hu et al. 2014).

Through DNA damage checkpoints, DDR mechanisms control if a cell can proceed through the cell cycle (Bartek and Lukas 2007). One key player in regulating the G1/S checkpoint in response to DNA damage is the tumor suppressor p53 (Senturk and Manfredi 2013). As explained before, in response to DNA damage, p53 can be phosphorylated by a variety of damage-induced kinases, such as the DDR kinase ATM. This activates p53's wide range of roles: activation of p21, leading to cell-cycle arrest; modulation of apoptosis through PUMA and NOXA activation; and upregulation of some DNA repair pathways, including NER and mismatch repair (Nakano and Vousden 2001; Shibue et al. 2003; Williams and Schumacher 2016; Ou and Schumacher 2018).

p53 has a short half-life; however, constant activation of DDR during aging leads to p53 stabilization. This constant p53 activation can promote cellular senescence if the cell cycle is irreversibly arrested (Ou and Schumacher 2018). This way, constant activation of p53 accelerates aging, since senescent cells affect the surrounding tissue microenvironment by preventing tissue homeostasis and remodeling and by promoting inflammation through SASP (Coppé et al. 2010).

Hematopoietic stem cells (HSCs) depend on their self-renewal capacity to be able to replenish the organism with blood and a functioning immune system (Chambers et al. 2007). Accumulation of oxidative DNA damage over time limits the rejuvenation of HSCs, in part due to persistent activation of both DDR and p53, thus triggering cellular senescence (Chambers et al. 2007; Behrens et al. 2014). This age-dependent decline in stem cell niche of the hematopoietic system has been shown to contribute to several age-related diseases, including cancer, neurodegeneration, and vascular disorders (Mangerich and Bürkle 2012). Persistent DNA damage was also described in different tissues from aged mice, including the liver, testis, kidney, lung, and brain, further explaining age-derived phenotypes (Sedelnikova et al. 2004).

Telomere shortening during replicative senescence was also shown to trigger persistent activation of DDR mechanisms and p53 (Hewitt et al. 2012). Telomeres consist of tandem TTAGGG repeats bound to proteins at chromosome termini, thus conferring protection to the chromosomal ends (Hewitt et al. 2012; Maynard et al. 2015). However, as organisms age, the ability to maintain telomere length decreases, leading to telomere shortening, which are recognized as DSB (Ou and Schumacher 2018).

Another source of permanent DNA damage stems from age-related mitochondrial dysfunction caused by accumulation of mitochondrial DNA damage over time. With aging, there is increased "leakage" of ROS from the mitochondria, leading to augmented DNA oxidation (Samper et al. 2003). This agrees with the oxidative stress theory of aging. One consequence of the increased oxidative damage caused by aged mitochondria is hyperactivation of poly(ADP-ribose)polymerase 1 (PARP1), a protein involved in the recruitment and repair of proteins by catalyzing the addition of PARylation chains (Wei and Yu 2016). Hyperactivation of PARP1 can be detrimental to the cells by increasing energy consumption. This causes NAD^+ imbalance and consequent SIRT1 deactivation, since both PARP1 and SIRT1 depend on NAD^+ for their activities (Fang et al. 2014). SIRT1 is a class III histone deacetylase and plays a role in many biological processes by influencing chromatin dynamics and stability through histone deacetylation, including DNA repair and telomere stability. Other functions of SIRT1 include autophagosome maturation, thus indorsing autophagy (Rahman and Islam 2011; Lin and Fang 2013). SIRT1 also promotes mitochondria functioning through deacetylation of the mitochondria regulator, PGC-1α (Ventura-Clapier et al. 2008). Furthermore, increased PARP1 activation can also promote cell death by parthanatos, a programmed cell death driven by the nuclear translocation of apoptosis-inducing factor (AIF) from the mitochondria. This cell death mechanism is independent on caspase activity and is promoted by PARylation (Kam et al. 2018; Wang and Ge 2020).

The process of aging is characterized by a series of cellular and molecular hallmarks that compromise cellular homeostasis and contribute to fitness decline associated with aging. As the organism ages, there are gradual changes in the epigenome that disrupt transcriptional silencing, shortening of telomeres, increase in cellular senescence, and accumulation of DNA damage over time. In turn, the declining of these mechanisms with time further accelerates the process of aging, contributing to increasing the chances of age-related diseases such as cancer and neurodegeneration (Created in BioRender).

4.6 Age-Related Physiological Changes and Nutritional Needs

Aging is accompanied by a gradual decline in physiological function. The rate of decline is different among individuals because of different factors such as genetics and lifestyle.

4.6.1 Some Physiological Changes

Age-related body composition changes include an increase in adipose tissue mass, especially visceral fat mass, and a decrease in lean muscle mass. Loss of muscle mass, strength, and function, also known as sarcopenia, may occur and impair the quality of life by contributing to functional decline, disability, frailty, and falls (Walston 2012). Decline in physical activity can even further accelerate sarcopenia (Montero-Fernández and Rexach-Serra 2013). Sufficient protein intake may preserve the declining muscle mass (Volkert et al. 2019). Body composition alterations also lead to a decrease in basal metabolic rate which should be met with suitable caloric intake (Deutz et al. 2014).

Aging is also associated with changes in kidney function. Glomerular filtration rate, which is used to assess kidney function, decreases by approximately 1.05 mL/min per year in individuals aged 70–110 years (Fehrman-Ekholm and Skeppholm 2004). Age-related changes in renal function may predispose the kidney to acute kidney injury, including normotensive ischemic nephropathy, as well as progressive chronic kidney disease (Weinstein and Anderson 2010).

Morphological and functional changes to the teeth occur throughout the lifetime as a consequence of the continuous exposure to chemical and mechanical forces (Carvalho and Lussi 2017). Suboptimal dental care, ill-fitting dentures, or neurodegenerative diseases cause chewing difficulties (Woo et al. 2018). Masticatory impairment and tooth loss are associated with increased intake of easy-to-chew foods that are high in sugar and decreased intake of high-fiber and vitamin-rich food (Kossioni 2018). Masticatory impairment is even associated with cognitive decline, including dementia. The poorer the mastication, the steeper was the cognitive decline in elders (Tada and Miura 2017).

There are age-related changes in the swallowing physiology (Humbert and Robbins 2008). Dysphagia may reduce food intake, especially proteins (Crary et al. 2012). This requires special dietary recommendations and modification of the food texture.

Age-related structural and functional changes in the brain have been recognized (Murman 2015). Prospective memory, source memory, and delayed free recall memory decline, while procedural, recognition, and temporal order memory remain the same (Harada et al. 2013). Aging is also a known risk factor for the development of neurodegenerative diseases, including Alzheimer's disease and Parkinson's disease (Hou et al. 2019).

4.6.2 Unhealthy Habits on ROS and Aging

Unhealthy practices like smoking may lead to genomic instability and telomere shortening. The dosage of cigarette smoking is negatively correlated with telomere length (Song et al. 2010). Valdes et al. conducted a study in WBCs of women and reported that telomeric DNA is lost at an average rate of "25.7–27.7 base pairs" per year, and with daily smoking of each pack of cigarettes, an additional "5 base pairs" are lost. The telomere attrition caused by smoking one pack of cigarettes a day over a period of 40 years is therefore equivalent to 7.4 years of life (Valdes et al. 2005). The molecular mechanism by which telomeres shrink in size could be explained by the oxidative damage induced by smoking and carcinogens found in cigarettes. For that reason, researchers have suggested antioxidant therapy to prevent oxidative stress (Babizhayev et al. 2011). Besides smoking, other unhealthy habits like alcohol consumption, unhealthy food, lack of exercise, obesity, etc. can lead to increased ROS production and other negative effects of aging. On the contrary, cellular defense mechanisms like ROS scavenging enzymes, antioxidants such as vitamins, flavonoids, and carotenoids can control ROS production and maintain telomeres integrity, reduce mitochondrial damage, and extend populations health span (Vidacek et al. 2017). People, therefore, can modulate their aging through comprehensive lifestyle changes.

4.6.3 Nutritional Needs for Older Adults

Older adults are encouraged to follow a Mediterranean diet that is high in vegetables, fruits, legumes, nuts, beans, cereals, grains, fish, and unsaturated fats such as olive oil. Diet rich in fish improves the cognition of older adults (Nurk et al. 2007). A Mediterranean diet may increase life span and slow down aging (Capurso et al. 2019). A cross-sectional study in the USA with 4676 subjects showed that greater adherence to the Mediterranean diet was associated with longer telomeres and that vegetable consumption is associated with longer telomeres, whereas cereal consumption is associated with shorter ones (Crous-Bou et al. 2019). Table 4.1 summarizes nutritional needs for older adults.

Table 4.1 Nutritional needs for older adults

Nutrient		Benefit(s)	Reference
Carbohydrates	Out of the recommended 30 kcal/kg body weight per day, 50–55% should be carbohydrates	• Provide energy	Volkert et al. (2019)
Fiber	20–25 per day	• Normalizes bowel functions	Volkert et al. (2019)
Protein	1 g/kg body weight per day Up to 2 g/kg body weight per day	• Preserve lean muscle mass and function • Promote recovery from illness, injury, or malnutrition	Volkert et al. (2019), Bauer et al. (2013)
Fat	Monosaturated fatty acids (avocados and olive oil) Polyunsaturated fatty acids (cold-water fatty fish and flaxseed)	• Omega-3 PUFA may help with age-related muscle wasting because they possess anti-inflammatory properties, enhance mTOR activation, and reduce insulin resistance • Omega-3 PUFA may have cardioprotective properties • Fish intake was associated with better cognitive performance	Nurk et al. (2007), Endo and Arita (2016), Dupont et al. (2019), Volkert et al. (2019)
Micronutrients	Folate (found in beef liver and fresh dark green leafy vegetables such as spinach) Vitamin B12 (found in beef liver and clams) Vitamin B6 (chickpeas and beef liver and salmon). Vitamin B2 (beef liver, beef, oats, and clams) Zinc (oysters and beef), magnesium (pumpkin and chia seeds), niacin (beef liver and chicken breast), vitamin C (citrus fruits), vitamin E (wheat germ oil, sunflower seeds, and almonds), magnesium (pumpkin seeds, chia seeds, and almonds) Calcium (yoghurt and sardines) along with vitamin D (trout and salmon)	• B vitamins help maintain genome stability • B vitamins are involved in homocysteine metabolism. Elevated homocysteine has been associated with higher risk for cardiovascular disease • Low B12 status was associated with brain atrophy and white matter damage • Help maintain genome stability • Maintain bone health	Kim et al. (2009), Smith and Refsum (2009), Fenech (2010), Debreceni and Debreceni (2014), Hartwig (2001), Fenech (2010), Sharif et al. (2012), Meehan and Penckofer (2014), Beto (2015)

4.7 Nutrigenomics and the Aging Process

As discussed before, oxidative stress has a significant effect on telomere shortening. Researchers have suggested that including antioxidants in one's diet can slow down the aging process by neutralizing free radicals in the cell (Sies 1997). Experiments were then conducted to assess the impact of antioxidant compounds found in food like vitamins, minerals, polyphenols, and omega-3 fatty acids in maintaining telomere length, therefore leading to healthy aging (García-Calzón et al. 2015). Both in vitro and in vivo studies have shown promising results in regard with antioxidants and telomere metabolism (Yokoo et al. 2004). Vitamins C, D, and E; folate; and β-carotene and the minerals zinc and magnesium have shown protection against oxidative stress and inflammation, sustaining telomere stability (Paul 2011). Van Aart et al. further investigated the role of vitamin D on telomere length. Outcomes indicated that vitamin D affects telomere length positively due to its anti-inflammatory properties. This study suggests that the higher vitamin D level, the longer the telomeres, highlighting the advantages of it on aging (Van Aart et al. 2018). There are nutrients that have the potential to conserve telomere length via mechanisms involved in cellular functions including inflammation, oxidative stress, DNA integrity, DNA methylation, and activity of telomerase.

Polyphenols are natural chemicals that are also thought to have a beneficial effect on telomere length and organismal aging. A study conducted on a Belgian population displayed antioxidant activities of theaflavins, polyphenols found in green and black tea along with a negative association with inflammatory markers (De Bacquer et al. 2006). In another study conducted on an elderly Chinese people, results showed that habitual tea drinkers have longer telomeres that contribute to an average increase of 5 years in life span compared to those who do not consume tea as much (Chan et al. 2010).

Another common naturally occurring polyphenol that is significantly correlated with telomere is resveratrol that activates SIRT1, an intracellular regulatory protein that controls important metabolic and physiological functions (Baur et al. 2006). Resveratrol is found in the skin of red grapes and has antioxidant and anti-inflammatory properties in addition to a beneficial effect on general health in mammals. It was also reported that resveratrol declines oxidative stress and attenuates inflammation and reduces the risk of age-related pathologies such as cardiovascular diseases and diabetes (Smoliga et al. 2011).

Over the last decade, omega-3 fatty acids have been recognized as critical molecules for well-being and a healthier cardiovascular system (Vidacek et al. 2017). Kiecolt-Glaser et al. showed in a randomized controlled 4-month trial that it is not omega-3 solely, but rather the ratio between omega-3 (n-3) and omega-6 (n-6) fatty acids that is crucial, as telomere length increases with declining n-6:n-3 plasma ratios regarding baseline telomere length (Kiecolt-Glaser et al. 2013). Omega-3-rich foods include walnuts, flaxseeds and flaxseed oil, tofu, canola oil, and fatty fish such as salmon, mackerel, anchovies, and sardines. Omega-6, however, are found in corn, soybeans, sunflower oils, meat, poultry, fish, and eggs.

Nutrition also plays an important role in epigenetics of aging and might explain variations in the aging pattern between identical twins (Fraga et al. 2005). The study of epigenetic variations during aging is a very attractive field since they can be modulated to extend longevity (Yi and Kim 2020). This can be done in an accessible way, for example, through diet changes (Pal and Tyler 2016). For example, caloric restriction (CR) was shown to be very effective in extending the life span of several organisms, from yeast to primates (Lee and Longo 2016). CR was shown to protect organisms against global loss of acetylation that occur during aging, by increasing the activation of class III HDAC, including yeast Sir2 and mammalian SIRT1 and SIRT6, promoting longevity (Ungvari et al. 2008; Vaquero and Reinberg 2009; Li et al. 2011; Peleg et al. 2016).

Another diet, the ketogenic diet (KD), is characterized by high-fat, low-protein, and low-carbohydrate intake. KD was shown to promote mitochondrial uncoupling protein activity, causing a reduction of ROS in the hippocampus of mice (Sullivan et al. 2004). Another study has demonstrated that high-fat diets rescue age-related neurological and metabolic phenotypes (Scheibye-Knudsen et al. 2014). Feeding CS mice models with a high-fat diet and supplementing with NAD^+ improved mitochondrial homeostasis by increasing the pool of acetyl-CoA, which is necessary for mitochondrial oxidative phosphorylation and augmented SIRT1 levels (Scheibye-Knudsen et al. 2014).

4.7.1 The Role of Micronutrients in Genome Stability

As mentioned in the previous sections of this chapter, genome instability has long been implicated as the main causal factor in aging. Vitamins and minerals are needed to maintain genome stability (Chatterjee 2001; Fenech 2002).

Vitamin B9 and other B vitamins are involved in DNA synthesis, DNA methylation, homocysteine metabolism, and glutathione synthesis. Dietary folic acid is first converted to dihydrofolate, then to tetrahydrofolate (THF), and finally to 5,10-methylenetetrahydrofolate (5,10-methylene-THF) by the vitamin B6-dependent serine hydroxymethyltransferase (George et al. 2019). The vitamin B2-dependent enzyme 5,10-methyltetrahydrofolate reductase (MTHFR) then catalyzes the irreversible reduction of 5,10-methylene-THF into 5-methyl-THF (Robien 2003; Jiang and Huang 2015). 5-methyl-THF then donates its methyl group to cobalamin converting it into methylcobalamin. Methylcobalamin then donates the methyl group to homocysteine to generate the essential amino acid methionine with the help of methionine synthase. Methionine can then be metabolized into the methylating agent S-adenosyl methionine (SAM), which is required in many biological processes such as to produce nucleic acids, proteins and neurotransmitters, and other methyltransferase reactions. After donating its methyl group, SAM becomes S-adenosyl homocysteine (SAH), an inhibitor of many transferases. SAH is catabolized by hydrolysis into adenosine and homocysteine. Homocysteine can also be converted into cysteine through the transsulfuration pathway in the presence of rate-limiting enzymes cystathionine-β synthase and cystathionine-γ lyase (Shane 2013; George et al. 2019). Cysteine is further converted into the scavenging antioxidant glutathione, which also takes part in nutrient metabolism and regulation of cellular events, including gene expression and apoptosis. Glutathione deficiency increases oxidative stress, which affects aging and the pathogenesis of many neurodegenerative diseases such as Alzheimer's disease and Parkinson's disease (Wu et al. 2004). Folate deficiency causes inadequate methylation of deoxyuridine monophosphate to deoxythymidine monophosphate, leading to chromosomal breaks (Blount et al. 1997). Folate deficiency also causes DNA hypomethylation and mitochondrial DNA deletions (Fenech 2012).

Niacin, another B vitamin, is required as a substrate for the enzyme poly (ADP-ribose)polymerase-1 which is involved in many DNA processes such as transcription, replication, chromatic remodeling, DNA repair, and telomere length maintenance (Morales et al. 2014). Niacin deficiency increased chromosome breaks and rearrangements (Fenech 2010). Vitamins with antioxidant activity such as vitamin C and vitamin E also play crucial roles in genomic stability as they prevent oxidation of DNA and lipids, and when they are deficient, there is an increase in the levels of DNA breaks and chromosome breaks (Fenech 2010).

As for minerals, magnesium plays an important role in genomic stability. It maintains membrane integrity and regulates cell proliferation, differentiation, and apoptosis (Hartwig 2001). Another important micronutrient for genomic stability, DNA integrity, and proper telomere function is zinc (Sharif et al. 2012). It is an essential cofactor for the antioxidant enzyme copper/zinc superoxide dismutase as well as a number of other zinc-finger proteins (Fenech 2002). Zinc deficiency increased DNA breaks and chromosomal damage rate (Fenech 2010).

References

Abugable AA et al (2019) DNA repair and neurological disease: From molecular understanding to the development of diagnostics and model organisms. DNA Repair 81:102669. https://doi.org/10.1016/j.dnarep.2019.102669

Acosta JC et al (2008) Chemokine signaling via the CXCR2 receptor reinforces senescence. Cell 133(6):1006–1018. https://doi.org/10.1016/j.cell.2008.03.038

Acosta JC et al (2013) A complex secretory program orchestrated by the inflammasome controls paracrine senescence. Nat Cell Biol 15(8):978–990. https://doi.org/10.1038/ncb2784

Agger K et al (2009) The H3K27me3 demethylase JMJD3 contributes to the activation of the INK4A-ARF locus in response to oncogene- and stress-induced senescence. Genes Dev 23(10): 1171–1176. https://doi.org/10.1101/gad.510809

Ahmed MS et al (2018) Hutchinson–gilford progeria syndrome: a premature aging disease. Mol Neurobiol 55(5):4417–4427. https://doi.org/10.1007/s12035-017-0610-7

Alves-Fernandes DK, Jasiulionis MG (2019) The role of SIRT1 on DNA damage response and epigenetic alterations in cancer. Int J Mol Sci 20(13):1–13. https://doi.org/10.3390/ijms20133153

Arancio W et al (2014) Epigenetic involvement in Hutchinson-Gilford progeria syndrome: a mini-review. Gerontology 60(3):197–203. https://doi.org/10.1159/000357206

Baar MP et al (2017) Targeted apoptosis of senescent cells restores tissue homeostasis in response to chemotoxicity and aging. Cell 169(1):132–147.e16. https://doi.org/10.1016/j.cell.2017.02.031

Babizhayev MA et al (2011) Telomere length is a biomarker of cumulative oxidative stress, biologic age, and an independent predictor of survival and therapeutic treatment requirement associated with smoking behavior. Am J Ther 18(6):e209–e226. https://doi.org/10.1097/MJT.0b013e3181cf8ebb

Baker DJ et al (2016) Naturally occurring p16(Ink4a)-positive cells shorten healthy lifespan. Nature 530(7589):184–189. https://doi.org/10.1038/nature16932

Baldensperger T et al (2020) Comprehensive analysis of posttranslational protein modifications in aging of subcellular compartments. Sci Rep 10(1):1–11. https://doi.org/10.1038/s41598-020-64265-0

Balestro E et al (2016) Immune inflammation and disease progression in idiopathic pulmonary fibrosis. PLoS One 11(5):e0154516–e0154516. https://doi.org/10.1371/journal.pone.0154516

Barradas M et al (2009) Histone demethylase JMJD3 contributes to epigenetic control of INK4a/ARF by oncogenic RAS. Genes Dev 23(10):1177–1182. https://doi.org/10.1101/gad.511109

Bartek J, Lukas J (2007) DNA damage checkpoints: from initiation to recovery or adaptation. Curr Opin Cell Biol 19(2):238–245. https://doi.org/10.1016/j.ceb.2007.02.009

Bauer J et al (2013) Evidence-based recommendations for optimal dietary protein intake in older people: a position paper from the PROT-AGE study group. J Am Med Dir Assoc 14(8). https://doi.org/10.1016/j.jamda.2013.05.021

Baumgart M et al (2014) RNA-seq of the aging brain in the short-lived fish N. furzeri—conserved pathways and novel genes associated with neurogenesis. Aging Cell 13(6):965–974. https://doi.org/10.1111/acel.12257

Baur JA et al (2006) Resveratrol improves health and survival of mice on a high-calorie diet. Nature 444(7117):337–342. https://doi.org/10.1038/nature05354

Bautista-Niño PK et al (2016) DNA damage: a main determinant of vascular aging. Int J Mol Sci 17(5):748. https://doi.org/10.3390/ijms17050748

Behrens A et al (2014) Impact of genomic damage and ageing on stem cell function. Nat Cell Biol 16(3):201–207. https://doi.org/10.1038/ncb2928

Benayoun BA, Pollina EA, Brunet A (2015) Epigenetic regulation of ageing: linking environmental inputs to genomic stability. Nat Rev Mol Cell Biol 16(10):593–610. https://doi.org/10.1038/nrm4048

Beto JA (2015) The role of calcium in human aging. Clin Nutr Res 4(1). https://doi.org/10.7762/cnr.2015.4.1.1

Blount BC et al (1997) Folate deficiency causes uracil misincorporation into human DNA and chromosome breakage: implications for cancer and neuronal damage. Proc Natl Acad Sci U S A 94(7). https://doi.org/10.1073/pnas.94.7.3290

Booth LN, Brunet A (2016) The aging epigenome. Mol Cell 62(5):728–744. https://doi.org/10.1016/j.molcel.2016.05.013

Bracken AP et al (2007) The Polycomb group proteins bind throughout the INK4A-ARF locus and are disassociated in senescent cells. Genes Dev 21(5):525–530. https://doi.org/10.1101/gad.415507

Brandt A, Krohne G, Großhans J (2008) The farnesylated nuclear proteins KUGELKERN and LAMIN B promote aging-like phenotypes in Drosophila flies. Aging Cell 7(4):541–551. https://doi.org/10.1111/j.1474-9726.2008.00406.x

Brouiette S et al (2003) White cell telomere length and risk of premature myocardial infarction. Arterioscler Thromb Vasc Biol 23(5):842–846. https://doi.org/10.1161/01.ATV.0000067426.96344.32

Brunt EM et al (2009) Portal chronic inflammation in nonalcoholic fatty liver disease (NAFLD): a histologic marker of advanced NAFLD-Clinicopathologic correlations from the nonalcoholic steatohepatitis clinical research network. Hepatology 49(3):809–820. https://doi.org/10.1002/hep.22724

Burd CE et al (2013) Monitoring tumorigenesis and senescence in vivo with a p16(INK4a)-luciferase model. Cell 152(1–2):340–351. https://doi.org/10.1016/j.cell.2012.12.010
Capurso C et al (2019) The Mediterranean diet slows down the progression of aging and helps to prevent the onset of frailty: a narrative review. Nutrients 12(1). https://doi.org/10.3390/nu12010035
Carvalho TS, Lussi A (2017) Age-related morphological, histological and functional changes in teeth. J Oral Rehabil. Blackwell Publishing Ltd:291–298. https://doi.org/10.1111/joor.12474
Cawthon RM et al (2003) Association between telomere length in blood and mortality in people aged 60 years or older. Lancet 361(9355):393–395. https://doi.org/10.1016/S0140-6736(03)12384-7
Chambers SM et al (2007) Aging hematopoietic stem cells decline in function and exhibit epigenetic dysregulation. PLoS Biol 5(8):1750–1762. https://doi.org/10.1371/journal.pbio.0050201
Chan R et al (2010) Chinese tea consumption is associated with longer telomere length in elderly Chinese men. Br J Nutr 103(1):107–113. https://doi.org/10.1017/S0007114509991383
Chatterjee M (2001) Vitamin D and genomic stability. Mutat Res Fundam Mol Mech Mutagen 475(1–2). https://doi.org/10.1016/S0027-5107(01)00080-X
Childs BG et al (2015) Cellular senescence in aging and age-related disease: from mechanisms to therapy. Nat Med 21(12):1424–1435. https://doi.org/10.1038/nm.4000
Cleaver JE et al (1999) A summary of mutations in the UV-sensitive disorders: xeroderma pigmentosum, Cockayne syndrome, and trichothiodystrophy. Hum Mutat 14(1):9–22. https://doi.org/10.1002/(SICI)1098-1004(1999)14:1<9::AID-HUMU2>3.0.CO;26
Coppé J-P et al (2006) Secretion of vascular endothelial growth factor by primary human fibroblasts at senescence. J Biol Chem 281(40):29568–29574. https://doi.org/10.1074/jbc.m603307200
Coppé J-P et al (2008) Senescence-associated secretory phenotypes reveal cell-nonautonomous functions of oncogenic RAS and the p 53 tumor suppressor. PLoS Biol 6(12):2853–2868. https://doi.org/10.1371/journal.pbio.0060301
Coppé J-P et al (2010) The senescence-associated secretory phenotype: the dark side of tumor suppression. Annu Rev Pathol 5:99–118. https://doi.org/10.1146/annurev-pathol-121808-102144
Crane MM et al (2019) DNA damage checkpoint activation impairs chromatin homeostasis and promotes mitotic catastrophe during aging. eLife 8:1–26. https://doi.org/10.7554/eLife.50778
Crary M et al (2012) Dysphagia in the elderly: management and nutritional considerations. Clin Interv Aging. https://doi.org/10.2147/CIA.S23404
Crous-Bou M, Molinuevo J-L, Sala-Vila A (2019) Plant-rich dietary patterns, plant foods and nutrients, and telomere length. Adv Nutr 10(Suppl_4). https://doi.org/10.1093/advances/nmz026
d'Adda di Fagagna F (2008) Living on a break: cellular senescence as a DNA-damage response. Nat Rev Cancer 8(7):512–522. https://doi.org/10.1038/nrc2440
Da Silva PFL, Schumacher B (2019) DNA damage responses in ageing. Open Biol 9(11). https://doi.org/10.1098/rsob.190168
Dang W et al (2009) Histone H4 lysine 16 acetylation regulates cellular lifespan. Nature 459(7248):802–807. https://doi.org/10.1038/nature08085
De Bacquer D et al (2006) Epidemiological evidence for an association between habitual tea consumption and markers of chronic inflammation. Atherosclerosis 189(2):428–435. https://doi.org/10.1016/j.atherosclerosis.2005.12.028
Debreceni B, Debreceni L (2014) The role of homocysteine-lowering B-vitamins in the primary prevention of cardiovascular disease. Cardiovasc Ther 32(3). https://doi.org/10.1111/1755-5922.12064
Demaria M et al (2017) Cellular senescence promotes adverse effects of chemotherapy and cancer relapse. Cancer Discov 7(2):165–176. https://doi.org/10.1158/2159-8290.CD-16-0241

Deutz NEP et al (2014) Protein intake and exercise for optimal muscle function with aging: recommendations from the ESPEN Expert Group. Clin Nutr 33(6):929–936. https://doi.org/10.1016/j.clnu.2014.04.007

Di Mitri D et al (2014) Tumour-infiltrating Gr-1+ myeloid cells antagonize senescence in cancer. Nature 515(7525):134–137. https://doi.org/10.1038/nature13638

Dinarello CA, Donath MY, Mandrup-Poulsen T (2010) Role of IL-1β in type 2 diabetes. Curr Opin Endocrinol Diabetes Obes:1. https://doi.org/10.1097/med.0b013e32833bf6dc

Dou Z et al (2017) Cytoplasmic chromatin triggers inflammation in senescence and cancer. Nature 550(7676):402–406. https://doi.org/10.1038/nature24050

Dupont J et al (2019) The role of omega-3 in the prevention and treatment of sarcopenia. Aging Clin Exp Res 31(6). https://doi.org/10.1007/s40520-019-01146-1

Eggert T et al (2016) Distinct functions of senescence-associated immune responses in liver tumor surveillance and tumor progression. Cancer Cell 30(4):533–547. https://doi.org/10.1016/j.ccell.2016.09.003

El-Khamisy SF (2011) To live or to die: a matter of processing damaged DNA termini in neurons. EMBO Mol Med. https://doi.org/10.1002/emmm.201000114

El-Khamisy SF, Caldecott KW (2006) TDP1-dependent DNA single-strand break repair and neurodegeneration. Mutagenesis 21(4):219–224. https://doi.org/10.1093/mutage/gel024

Endo J, Arita M (2016) Cardioprotective mechanism of omega-3 polyunsaturated fatty acids. J Cardiol 67(1). https://doi.org/10.1016/j.jjcc.2015.08.002

Eyman D et al (2009) CCL5 secreted by senescent aged fibroblasts induces proliferation of prostate epithelial cells and expression of genes that modulate angiogenesis. J Cell Physiol 220(2): 376–381. https://doi.org/10.1002/jcp.21776

Fang EF et al (2014) Defective mitophagy in XPA via PARP-1 hyperactivation and NAD +/SIRT1 reduction. Cell 157(4):882–896. https://doi.org/10.1016/j.cell.2014.03.026

Fehrman-Ekholm I, Skeppholm L (2004) Renal function in the elderly (>70 years old) measured by means of iohexol clearance, serum creatinine, serum urea and estimated clearance. Scand J Urol Nephrol 38(1). https://doi.org/10.1080/00365590310015750

Fenech M (2002) Micronutrients and genomic stability: a new paradigm for recommended dietary allowances (RDAs). Food Chem Toxicol 40(8). https://doi.org/10.1016/S0278-6915(02)00028-5

Fenech MF (2010) Nutriomes and nutrient arrays—the key to personalised nutrition for DNA damage prevention and cancer growth control. Genome Integr 1(1). https://doi.org/10.1186/2041-9414-1-11

Fenech M (2012) Folate (vitamin B9) and vitamin B12 and their function in the maintenance of nuclear and mitochondrial genome integrity. Mutat Res Fundam Mol Mech Mutagen 733(1–2). https://doi.org/10.1016/j.mrfmmm.2011.11.003

Fenech M et al (2011) Nutrigenetics and nutrigenomics: viewpoints on the current status and applications in nutrition research and practice. J Nutrigenet Nutrigenomics:69–89. https://doi.org/10.1159/000327772

Feser J et al (2010) Elevated histone expression promotes life span extension. Mol Cell 39(5): 724–735. https://doi.org/10.1016/j.molcel.2010.08.015

Fraga MF et al (2005) Epigenetic differences arise during the lifetime of monozygotic twins. Proc Natl Acad Sci U S A 102(30):10604–10609. https://doi.org/10.1073/pnas.0500398102

Franceschi C, Campisi J (2014) Chronic inflammation (inflammaging) and its potential contribution to age-associated diseases. J Gerontol A Biol Sci Med Sci 69(Suppl. 1):S4–S9. https://doi.org/10.1093/gerona/glu057

Fumagalli M et al (2012) Telomeric DNA damage is irreparable and causes persistent DNA-damage-response activation. Nat Cell Biol 14(4):355–365. https://doi.org/10.1038/ncb2466

García-Calzón S et al (2015) Dietary total antioxidant capacity is associated with leukocyte telomere length in a children and adolescent population. Clin Nutr 34(4):694–699. https://doi.org/10.1016/j.clnu.2014.07.015

George AK et al (2019) Hydrogen sulfide intervention in cystathionine-β-synthase mutant mouse helps restore ocular homeostasis. Int J Ophthalmol 12(5):754–764. https://doi.org/10.18240/ijo.2019.05.09

Gil J, Peters G (2006) Regulation of the INK4b–ARF–INK4a tumour suppressor locus: all for one or one for all. Nat Rev Mol Cell Biol 7(9):667–677. https://doi.org/10.1038/nrm1987

Glück S et al (2017) Innate immune sensing of cytosolic chromatin fragments through cGAS promotes senescence. Nat Cell Biol 19(9):1061–1070. https://doi.org/10.1038/ncb3586

Gorbunova V, Seluanov A (2016) DNA double strand break repair, aging and the chromatin connection. Mutat Res Fundam Mol Mechan Mutagen 788:2–6. https://doi.org/10.1016/j.mrfmmm.2016.02.004

Haithcock E et al (2005) Age-related changes of nuclear architecture in Caenorhabditis elegans. Proc Natl Acad Sci U S A 102(46):16690–16695. https://doi.org/10.1073/pnas.0506955102

Hanahan D, Weinberg RA (2011) Hallmarks of cancer: the next generation. Cell 144(5):646–674. https://doi.org/10.1016/j.cell.2011.02.013

Harada CN, Natelson Love MC, Triebel KL (2013) Normal cognitive aging. Clin Geriatr Med 29(4). https://doi.org/10.1016/j.cger.2013.07.002

Harris SL, Levine AJ (2005) The p53 pathway: positive and negative feedback loops. Oncogene 24(17):2899–2908. https://doi.org/10.1038/sj.onc.1208615

Hartwig A (2001) Role of magnesium in genomic stability. Mutat Res Fundam Mol Mech Mutagen 475(1–2). https://doi.org/10.1016/S0027-5107(01)00074-4

Hayflick L (1965) The limited in vitro lifetime of human diploid cell strains. Exp Cell Res 37(3): 614–636. https://doi.org/10.1016/0014-4827(65)90211-9

Helman A et al (2016) p16(Ink4a)-induced senescence of pancreatic beta cells enhances insulin secretion. Nat Med 22(4):412–420. https://doi.org/10.1038/nm.4054

Hernandez-Segura A et al (2017) Unmasking transcriptional heterogeneity in senescent cells. Curr Biol 27(17):2652–2660.e4. https://doi.org/10.1016/j.cub.2017.07.033

Herranz N et al (2015) mTOR regulates MAPKAPK2 translation to control the senescence-associated secretory phenotype. Nat Cell Biol 17(9):1205–1217. https://doi.org/10.1038/ncb3225

Hewitt G et al (2012) Telomeres are favoured targets of a persistent DNA damage response in ageing and stress-induced senescence. Nat Commun 3. https://doi.org/10.1038/ncomms1708

Hou Y et al (2019) Ageing as a risk factor for neurodegenerative disease. Nat Rev Neurol 15(10). https://doi.org/10.1038/s41582-019-0244-7

Houtkooper RH, Pirinen E, Auwerx J (2012) Sirtuins as regulators of metabolism and healthspan. Nat Rev Mol Cell Biol 13(4):225–238. https://doi.org/10.1038/nrm3293

Hu Z et al (2014) Nucleosome loss leads to global transcriptional up-regulation and genomic instability during yeast aging. Genes Dev 28(4):396–408. https://doi.org/10.1101/gad.233221.113

Humbert IA, Robbins J (2008) Dysphagia in the elderly. Phys Med Rehabil Clin N Am 19(4). https://doi.org/10.1016/j.pmr.2008.06.002

Ivanov A et al (2013) Lysosome-mediated processing of chromatin in senescence. J Cell Biol 202(1):129–143. https://doi.org/10.1083/jcb.201212110

Janzen V et al (2006) Stem-cell ageing modified by the cyclin-dependent kinase inhibitor p16INK4a. Nature 443(7110):421–426. https://doi.org/10.1038/nature05159

Jeck WR, Siebold AP, Sharpless NE (2012) Review: a meta-analysis of GWAS and age-associated diseases. Aging Cell 11(5):727–731. https://doi.org/10.1111/j.1474-9726.2012.00871.x

Jeon OH et al (2017) Local clearance of senescent cells attenuates the development of post-traumatic osteoarthritis and creates a pro-regenerative environment. Nat Med 23(6):775–781. https://doi.org/10.1038/nm.4324

Jiang Y, Huang C (2015) Polymorphism and cognition. In: Diet and nutrition in dementia and cognitive decline. Elsevier. https://doi.org/10.1016/B978-0-12-407824-6.00032-X

Jung M, Pfeifer GP (2015) Aging and DNA methylation. BMC Biol 13(1):1–8. https://doi.org/10.1186/S12915-015-0118-4

Kaeberlein M, McVey M, Guarente L (1999) The SIR2/3/4 complex and SIR2 alone promote longevity in Saccharomyces cerevisiae by two different mechanisms. *Genes Dev* 13(19): 2570–2580. https://doi.org/10.1101/gad.13.19.2570
Kam TI et al (2018) Poly(ADP-ribose) drives pathologic a-synuclein neurodegeneration in Parkinson's disease. Science 362(6414). https://doi.org/10.1126/science.aat8407
Kane AE, Sinclair DA (2019) Epigenetic changes during aging and their reprogramming potential. Crit Rev Biochem Mol Biol 54(1):61–83. https://doi.org/10.1080/10409238.2019.1570075
Kang T-W et al (2011) Senescence surveillance of pre-malignant hepatocytes limits liver cancer development. Nature 479(7374):547–551. https://doi.org/10.1038/nature10599
Kang C et al (2015) The DNA damage response induces inflammation and senescence by inhibiting autophagy of GATA4. Science 349(6255):aaa5612–aaa5612. https://doi.org/10.1126/science.aaa5612
Katyal S et al (2007) TDP1 facilitates chromosomal single-strand break repair in neurons and is neuroprotective in vivo. EMBO J 26(22):4720–4731. https://doi.org/10.1038/sj.emboj.7601869
Kiecolt-Glaser JK et al (2013) Omega-3 fatty acids, oxidative stress, and leukocyte telomere length: a randomized controlled trial. Brain Behav Immun 28:16–24. https://doi.org/10.1016/j.bbi.2012.09.004
Kim WY, Sharpless NE (2006) The regulation of INK4/ARF in cancer and aging. Cell 127(2): 265–275. https://doi.org/10.1016/j.cell.2006.10.003
Kim K, Friso S, Choi S-W (2009) DNA methylation, an epigenetic mechanism connecting folate to healthy embryonic development and aging. J Nutr Biochem 20(12). https://doi.org/10.1016/j.jnutbio.2009.06.008
Kim YH et al (2017) Senescent tumor cells lead the collective invasion in thyroid cancer. Nat Commun 8:15208. https://doi.org/10.1038/ncomms15208
Kossioni AE (2018) The association of poor oral health parameters with malnutrition in older adults: a review considering the potential implications for cognitive impairment. Nutrients 10(11). https://doi.org/10.3390/NU10111709
Krishnamurthy J et al (2004) Ink4a/Arf expression is a biomarker of aging. J Clin Invest 114(9): 1299–1307. https://doi.org/10.1172/JCI22475
Krtolica A et al (2001) Senescent fibroblasts promote epithelial cell growth and tumorigenesis: a link between cancer and aging. Proc Natl Acad Sci U S A 98(21):12072–12077. https://doi.org/10.1073/pnas.211053698
Kruiswijk F, Labuschagne CF, Vousden KH (2015) p53 in survival, death and metabolic health: a lifeguard with a licence to kill. Nat Rev Mol Cell Biol 16(7):393–405. https://doi.org/10.1038/nrm4007
Kuilman T, Peeper DS (2009) Senescence-messaging secretome: SMS-ing cellular stress. Nat Rev Cancer 9(2):81–94. https://doi.org/10.1038/nrc2560
Kuilman T et al (2008) Oncogene-induced senescence relayed by an interleukin-dependent inflammatory network. Cell 133(6):1019–1031. https://doi.org/10.1016/j.cell.2008.03.039
Kuilman T et al (2010) The essence of senescence. Genes Dev 24(22):2463–2479. https://doi.org/10.1101/gad.1971610
Laberge R-M et al (2015) MTOR regulates the pro-tumorigenic senescence-associated secretory phenotype by promoting IL1A translation. Nat Cell Biol 17(8):1049–1061. https://doi.org/10.1038/ncb3195
Larson K et al (2012) Heterochromatin formation promotes longevity and represses ribosomal RNA synthesis. PLoS Genet 8(1). https://doi.org/10.1371/journal.pgen.1002473
Lawrence M, Daujat S, Schneider R (2016) Lateral thinking: how histone modifications regulate gene expression. Trends Genet 32(1):42–56. https://doi.org/10.1016/j.tig.2015.10.007
Lee C, Longo V (2016) Dietary restriction with and without caloric restriction for healthy aging. F1000Res 5(0):1–7. https://doi.org/10.12688/f1000research.7136.1
Lee JH et al (2020) Heterochromatin: an epigenetic point of view in aging. Exp Mol Med 52(9): 1466–1474. https://doi.org/10.1038/s12276-020-00497-4

Li Y, Daniel M, Tollefsbol TO (2011) Epigenetic regulation of caloric restriction in aging. BMC Med 9:1–12. https://doi.org/10.1186/1741-7015-9-98

Li S et al (2017) Identification, replication and characterization of epigenetic remodelling in the aging genome: a cross population analysis. Sci Rep 7(1). https://doi.org/10.1038/s41598-017-08346-7

Li X et al (2018) Regulation of chromatin and gene expression by metabolic enzymes and metabolites. Nat Rev Mol Cell Biol 19(9):563–578. https://doi.org/10.1038/s41580-018-0029-7

Liao C et al (2018) Chromosomal break repair by controlling TDP1 article UCHL3 regulates topoisomerase-induced chromosomal break repair by controlling TDP1 proteostasis. Cell Rep 23(11):3352–3365. https://doi.org/10.1016/j.celrep.2018.05.033

Libby P, Ridker PM, Maseri A (2002) Inflammation and atherosclerosis. Circulation 105(9): 1135–1143. https://doi.org/10.1161/hc0902.104353

Lieber MR, Karanjawala ZE (2004) Ageing, repetitive genomes and DNA damage. Nat Rev Mol Cell Biol 5(1):69–75. https://doi.org/10.1038/nrm1281

Lin Z, Fang D (2013) The roles of SIRT1 in cancer. Genes Cancer 4(3–4):97–104. https://doi.org/10.1177/1947601912475079

Lomas NJ et al (2012) Idiopathic pulmonary fibrosis: immunohistochemical analysis provides fresh insights into lung tissue remodelling with implications for novel prognostic markers. Int J Clin Exp Pathol 5(1):58–71

López-Otín C et al (2013) The hallmarks of aging. Cell 153(6):1194–1217. https://doi.org/10.1016/j.cell.2013.05.039

Lyu G et al (2018) TGF-β signaling alters H4K20me3 status via miR-29 and contributes to cellular senescence and cardiac aging. Nat Commun 9(1):1–13. https://doi.org/10.1038/s41467-018-04994-z

Machida S et al (2018) Structural basis of heterochromatin formation by human HP1. Mol Cell 69(3):385–397.e8. https://doi.org/10.1016/j.molcel.2017.12.011

Madabhushi R, Pan L, Tsai LH (2014) DNA damage and its links to neurodegeneration. Neuron 83(2):266–282. https://doi.org/10.1016/j.neuron.2014.06.034

Maison C, Almouzni G (2004) HP1 and the dynamics of heterochromatin maintenance. Nat Rev Mol Cell Biol 5(4):296–304. https://doi.org/10.1038/nrm1355

Mangerich A, Bürkle A (2012) Pleiotropic cellular functions of PARP1 in longevity and aging: genome maintenance meets inflammation. Oxid Med Cell Longev. https://doi.org/10.1155/2012/321653

Maskey RS et al (2014) Spartan deficiency causes genomic instability and progeroid phenotypes. Nat Commun 5. https://doi.org/10.1038/ncomms6744

Matheu A et al (2009) Anti-aging activity of theInk4/Arflocus. Aging Cell 8(2):152–161. https://doi.org/10.1111/j.1474-9726.2009.00458.x

Maxwell PH (2016) What might retrotransposons teach us about aging? Curr Genet 62(2):277–282. https://doi.org/10.1007/s00294-015-0538-2

Maxwell PH, Burhans WC, Curcio MJ (2011) Retrotransposition is associated with genome instability during chronological aging. Proc Natl Acad Sci U S A 108(51):20376–20381. https://doi.org/10.1073/pnas.1100271108

Maynard S et al (2015) DNA damage, DNA repair, aging, and neurodegeneration. Cold Spring Harb Perspect Biol:1–18

McHugh D, Gil J (2018) Senescence and aging: causes, consequences, and therapeutic avenues. J Cell Biol 217(1):65–77. https://doi.org/10.1083/jcb.201708092

Meehan M, Penckofer S (2014) The role of vitamin D in the aging adult. J Aging Gerontol 2(2). https://doi.org/10.12974/2309-6128.2014.02.02.1

Melk A et al (2004) Expression of p16INK4a and other cell cycle regulator and senescence associated genes in aging human kidney. Kidney Int 65(2):510–520. https://doi.org/10.1111/j.1523-1755.2004.00438.x

Mitchell SE et al (2016) The effects of graded levels of calorie restriction: V. Impact of short term calorie and protein restriction on physical activity in the C57BL/6 mouse. Oncotarget 7(15): 19147–19170. https://doi.org/10.18632/oncotarget.8158
Montero-Fernández N, Rexach-Serra JA (2013) Role of exercise on sarcopenia in the elderly. Eur J Phys Rehabil Med:131–143
Morales J et al (2014) Review of poly (ADP-ribose) polymerase (PARP) mechanisms of action and rationale for targeting in cancer and other diseases. Crit Rev Eukaryot Gene Expr 24(1). https://doi.org/10.1615/CritRevEukaryotGeneExpr.2013006875
Moskalev AA et al (2013) The role of DNA damage and repair in aging through the prism of Koch-like criteria. Ageing Res Rev 12(2):661–684. https://doi.org/10.1016/j.arr.2012.02.001
Muñoz-Espín D, Serrano M (2014) Cellular senescence: from physiology to pathology. Nat Rev Mol Cell Biol 15(7):482–496. https://doi.org/10.1038/nrm3823
Muñoz-Espín D et al (2013) Programmed cell senescence during mammalian embryonic development. Cell 155(5):1104–1118. https://doi.org/10.1016/j.cell.2013.10.019
Murman D (2015) The impact of age on cognition. Semin Hear 36(03). https://doi.org/10.1055/s-0035-1555115
Musich PR, Zou Y (2011) DNA-damage accumulation and replicative arrest in Hutchinson-Gilford progeria syndrome. Biochem Soc Trans 39(6):1764–1769. https://doi.org/10.1042/BST20110687
Nakano K, Vousden KH (2001) PUMA, a novel proapoptotic gene, is induced by p53. Mol Cell 7(3):683–694. https://doi.org/10.1016/S1097-2765(01)00214-3
Narita M et al (2006) A novel role for high-mobility group A proteins in cellular senescence and heterochromatin formation. Cell 126(3):503–514. https://doi.org/10.1016/j.cell.2006.05.052
Ni Z et al (2012) Two SET domain containing genes link epigenetic changes and aging in Caenorhabditis elegans. Aging Cell 11(2):315–325. https://doi.org/10.1111/j.1474-9726.2011.00785.x
Niccoli T, Partridge L (2012) Ageing as a risk factor for disease. Curr Biol 22(17):R741–R752. https://doi.org/10.1016/j.cub.2012.07.024
Nicetto D, Zaret KS (2019) Role of H3K9me3 heterochromatin in cell identity establishment and maintenance. Curr Opin Genet Dev 55:1–10. https://doi.org/10.1016/j.gde.2019.04.013
Nicetto D et al (2019) H3K9me3-heterochromatin loss at protein-coding genes enables developmental lineage specification. Science 363(6424):294–297. https://doi.org/10.1126/science.aau0583
Niedernhofer LJ et al (2018) Nuclear genomic instability and aging. Annu Rev Biochem 87:295–322. https://doi.org/10.1146/annurev-biochem-062917-012239
Nurk E et al (2007) Cognitive performance among the elderly and dietary fish intake: the Hordaland Health Study. Am J Clin Nutr 86(5). https://doi.org/10.1093/ajcn/86.5.1470
O'Sullivan RJ, Karlseder J (2012) The great unravelling: chromatin as a modulator of the aging process. Trends Biochem Sci 37(11):466–476. https://doi.org/10.1016/j.tibs.2012.08.001
Oshima J, Sidorova JM, Monnat RJ Jr (2017) Werner syndrome: clinical features, pathogenesis and potential therapeutic interventions. Ageing Res Rev 33:105–114. https://doi.org/10.1016/j.arr.2016.03.002
Ou HL, Schumacher B (2018) DNA damage responses and p53 in the aging process. Blood 131(5):488–495. https://doi.org/10.1182/blood-2017-07-746396
Pal S, Tyler JK (2016) Epigenetics and aging. Sci Adv 2(July). https://doi.org/10.1007/978-3-319-55530-0_123
Passos JF et al (2007) Mitochondrial dysfunction accounts for the stochastic heterogeneity in telomere-dependent senescence. PLoS Biol 5(5):e110. https://doi.org/10.1371/journal.pbio.0050110
Paul L (2011) Diet, nutrition and telomere length. J Nutr Biochem 22(10):895–901. https://doi.org/10.1016/j.jnutbio.2010.12.001
Pazolli E et al (2009) Senescent stromal-derived osteopontin promotes preneoplastic cell growth. Cancer Res 69(3):1230–1239. https://doi.org/10.1158/0008-5472.CAN-08-2970

Peleg S et al (2016) The metabolic impact on histone acetylation and transcription in ageing. Trends Biochem Sci 41(8):700–711. https://doi.org/10.1016/j.tibs.2016.05.008

Pérez RF et al (2018) Distinct chromatin signatures of DNA hypomethylation in aging and cancer. Aging Cell 17(3). https://doi.org/10.1111/acel.12744

Petr MA et al (2020) Protecting the aging genome. Trends Cell Biol 30(2):117–132. https://doi.org/10.1016/j.tcb.2019.12.001

Rahman S, Islam R (2011) Mammalian Sirt1: insights on its biological functions. Cell Commun Signal 9(1):11. https://doi.org/10.1186/1478-811X-9-11

Robien K (2003) 5,10-Methylenetetrahydrofolate reductase polymorphisms and leukemia risk: a HuGE minireview. Am J Epidemiol 157(7). https://doi.org/10.1093/aje/kwg024

Rodier F et al (2011) DNA-SCARS: distinct nuclear structures that sustain damage-induced senescence growth arrest and inflammatory cytokine secretion. J Cell Sci 124(Pt. 1):68–81. https://doi.org/10.1242/jcs.071340

Sadaie M et al (2013) Redistribution of the Lamin B1 genomic binding profile affects rearrangement of heterochromatic domains and SAHF formation during senescence. Genes Dev 27(16):1800–1808. https://doi.org/10.1101/gad.217281.113

Salama R et al (2014) Cellular senescence and its effector programs. Genes Dev 28(2):99–114. https://doi.org/10.1101/gad.235184.113

Samper E, Nicholls DG, Melov S (2003) Mitochondrial oxidative stress causes chromosomal instability of mouse embryonic fibroblasts. Aging Cell 2(5):277–285. https://doi.org/10.1046/j.1474-9728.2003.00062.x

Saxton RA, Sabatini DM (2017) mTOR signaling in growth, metabolism, and disease. Cell 168(6):960–976. https://doi.org/10.1016/j.cell.2017.02.004

Scaffidi P, Misteli T (2006) Lamin A-dependent nuclear defects in human aging. Science 312 (May):1059–1064

Scheibye-Knudsen M et al (2014) A high-fat diet and NAD+ activate sirt1 to rescue premature aging in cockayne syndrome. Cell Metab 20(5):840–855. https://doi.org/10.1016/j.cmet.2014.10.005

Sedelnikova OA et al (2004) Senescing human cells and ageing mice accumulate DNA lesions with unrepairable double-strand breaks. Nat Cell Biol 6(2):168–170. https://doi.org/10.1038/ncb1095

Sen P et al (2016) Epigenetic mechanisms of longevity and aging. Cell 166(4):822–839. https://doi.org/10.1016/j.cell.2016.07.050

Senturk E, Manfredi JJ (2013) p53 and cell cycle effects after DNA damage. Methods Mol Biol 962(1):49–61. https://doi.org/10.1007/978-1-62703-236-0_4

Shammas MA (2011) Telomeres, lifestyle, cancer, and aging. Curr Opin Clin Nutr Metab Care 14(1):28–34. https://doi.org/10.1097/MCO.0B013E32834121B1

Shane B (2013) Folate and vitamin B12 function. In: Encyclopedia of biological chemistry. Elsevier. https://doi.org/10.1016/B978-0-12-378630-2.00039-6

Sharif R et al (2012) The role of zinc in genomic stability. Mutat Res Fundam Mol Mech Mutagen 733(1–2). https://doi.org/10.1016/j.mrfmmm.2011.08.009

Sharpless NE, Sherr CJ (2015) Forging a signature of in vivo senescence. Nat Rev Cancer 15(7):397–408. https://doi.org/10.1038/nrc3960

Shibue T et al (2003) Integral role of Noxa in p53-mediated apoptotic response. Genes Dev 17(18):2233–2238. https://doi.org/10.1101/gad.1103603

Shumaker DK et al (2006) Mutant nuclear lamin A leads to progressive alterations of epigenetic control in premature aging. Proc Natl Acad Sci U S A 103(23):8703–8708. https://doi.org/10.1073/pnas.0602569103

Sies H (1997) Oxidative stress: oxidants and antioxidants. Exp Physiol 82(2):291–295. https://doi.org/10.1113/expphysiol.1997.sp004024

Smith AD, Refsum H (2009) Vitamin B-12 and cognition in the elderly. Am J Clin Nutr 89(2). https://doi.org/10.3945/ajcn.2008.26947D

Smoliga JM, Baur JA, Hausenblas HA (2011) Resveratrol and health—a comprehensive review of human clinical trials. Mol Nutr Food Res 55(8):1129–1141. https://doi.org/10.1002/mnfr.201100143

Solomon JM et al (2006) Inhibition of SIRT1 catalytic activity increases p53 acetylation but does not alter cell survival following DNA damage. Mol Cell Biol 26(1):28–38. https://doi.org/10.1128/MCB.26.1.28-38.2006

Song Z et al (2010) Lifestyle impacts on the aging-associated expression of biomarkers of DNA damage and telomere dysfunction in human blood. Aging Cell 9(4):607–615. https://doi.org/10.1111/j.1474-9726.2010.00583.x

Song S et al (2020) Senescent cells: emerging targets for human aging and age-related diseases. Trends Biochem Sci 45(7):578–592. https://doi.org/10.1016/j.tibs.2020.03.008

Srinivas N, Rachakonda S, Kumar R (2020) Telomeres and telomere length: a general overview. Cancers 12(3):558. https://doi.org/10.3390/CANCERS12030558

Stegeman R, Weake VM (2017) Transcriptional signatures of aging. J Mol Biol 429(16): 2427–2437. https://doi.org/10.1016/j.jmb.2017.06.019

Storer M et al (2013) Senescence is a developmental mechanism that contributes to embryonic growth and patterning. Cell 155(5):1119–1130. https://doi.org/10.1016/j.cell.2013.10.041

Sullivan PG et al (2004) The ketogenic diet increases mitochondrial uncoupling protein levels and activity. Ann Neurol 55(4):576–580. https://doi.org/10.1002/ana.20062

Tada A, Miura H (2017) Association between mastication and cognitive status: a systematic review. Arch Gerontol Geriatr 70:44–53. https://doi.org/10.1016/J.ARCHGER.2016.12.006

Takahashi A et al (2018) Downregulation of cytoplasmic DNases is implicated in cytoplasmic DNA accumulation and SASP in senescent cells. Nat Commun 9(1):1249. https://doi.org/10.1038/s41467-018-03555-8

Tan L et al (2017) Naked mole rat cells have a stable epigenome that resists iPSC reprogramming. Stem Cell Reports 9(5):1721–1734. https://doi.org/10.1016/j.stemcr.2017.10.001

Tasdemir N et al (2016) BRD4 connects enhancer remodeling to senescence immune surveillance. Cancer Discov 6(6):612–629. https://doi.org/10.1158/2159-8290.CD-16-0217

Tasselli L et al (2016) SIRT6 deacetylates H3K18ac at pericentric chromatin to prevent mitotic errors and cellular senescence. Nat Struct Mol Biol 23(5):434–**440**. https://doi.org/10.1038/nsmb.3202

Tasselli L, Zheng W, Chua KF (2017) SIRT6: novel mechanisms and links to aging and disease. Trends Endocrinol Metab 28(3):168–185. https://doi.org/10.1016/j.tem.2016.10.002

Tilstra JS et al (2012) NF-κB inhibition delays DNA damage-induced senescence and aging in mice. J Clin Investig 122(7):2601–2612. https://doi.org/10.1172/JCI45785

Tsurumi A, Li WX (2012) Global heterochromatin loss: a unifying theory of aging? Epigenetics 7(7):680–688. https://doi.org/10.4161/epi.20540

Ungvari Z et al (2008) Mechanisms underlying caloric restriction and lifespan regulation: implications for vascular aging. Circ Res 102(5):519–529. https://doi.org/10.1161/CIRCRESAHA.107.168369

Valdes AM et al (2005) Obesity, cigarette smoking, and telomere length in women. Lancet 366(9486):662–664. https://doi.org/10.1016/S0140-6736(05)66630-5

Van Aart C et al (2018) Vitamin D as predictor of telomere length in the transition from child to adolescent. Rev Epidemiol Sante Publique 66:S237. https://doi.org/10.1016/j.respe.2018.05.015

Van Meter M et al (2014) SIRT6 represses LINE1 retrotransposons by ribosylating KAP1 but this repression fails with stress and age. Nat Commun 5. https://doi.org/10.1038/ncomms6011

Vaquero A, Reinberg D (2009) Calorie restriction and the exercise of chromatin. Genes Dev 23(16): 1849–1869. https://doi.org/10.1101/gad.1807009

Vaz B et al (2016) Metalloprotease SPRTN/DVC1 orchestrates replication-coupled DNA-protein crosslink repair. Mol Cell 64(4):704–719. https://doi.org/10.1016/j.molcel.2016.09.032

Ventura-Clapier R, Garnier A, Veksler V (2008) Transcriptional control of mitochondrial biogenesis: the central role of PGC-1α. *Cardiovasc Res* 79(2):208–217. https://doi.org/10.1093/cvr/cvn098

Vidacek NŠ et al (2017) Telomeres, nutrition, and longevity: can we really navigate our aging? J Gerontol A Biol Sci Med Sci 73(1):39–47. https://doi.org/10.1093/gerona/glx082

5.2 How Mental Health Affects Nutrition

Lifestyles, including diet, have proven to at least partially impact mood through various pathways (Lopresti et al. 2013). Besides eating disorders and appetite-related symptoms related to mental illness such as depression, patients with mental health problems may have other challenges that influence their nutritional status such as poor eating habits, low motivation to prepare meals, consumption of unhealthy foods, or eating junk foods (Teasdale et al. 2017) which all increase the likelihood of adopting unhealthy dietary habits (van Gool et al. 2003). Ironically, this means that affective states in themselves can influence poor food choices, which creates a vicious cycle. For example, high consumption of processed carbohydrates is believed to increase risk of depression (Firth et al. 2020), and depression has been shown to be a risk factor for metabolic syndrome (Crichton et al. 2016). This overlap between appetite and mood can be explained by the fact that the same neurotransmitters might be involved in both such as serotonin (Bremner et al. 2020). In terms of comorbidity, the co-occurrence of mental health problems and eating-related conditions is intriguing. For example, patients with depression and anxiety are more likely to display GIT symptoms than healthy controls (Dinan and Quigley 2011). One study also noted that metabolic syndrome, rather than obesity or diet, is correlated with depression prognosis (García-Toro et al. 2016).

Metabolic abnormalities in psychiatric disorders are not uncommon. For example, there is high comorbidity between psychosis and type 2 diabetes mellitus which has been attributed to different factors including the use of antipsychotics and secondary weight gain, impaired lifestyle and nutrition, and genetic susceptibility (Mizuki et al. 2021). Additionally, some specific psychotropic medications, such as olanzapine and mirtazapine, known for their therapeutic use in affective disorders are also known for their significant effect on increasing appetite and weight gain (Alam et al. 2013). Such findings do not only give insight into the interaction between psychiatric and metabolic disorders but also the need for dietary advice before the initiation of treatment.

5.3 Nutrition and Mental Health

Research evidence suggests a link between specific nutritional deficiencies and mental health outcomes as will be discussed later in this chapter. Although no causal relationships have yet been established (Firth et al. 2020), a strong association is emerging between diet and psychological states. Viewing mental health from a biopsychosocial perspective, it is important to shed light on nutrition as an essential element in the biopsychosocial triad of functioning. With a global transition to foods that are high in caloric values, but low in nutritional values (Kachani and de Lima Furtado 2020), more evidence for an effect of food on mood has started to emerge. On the least healthy end of the continuum, Western style diets have been associated with increased depressive symptoms (Altun et al. 2019) and high intake of processed foods has been correlated with higher levels of anxiety (Jacka et al. 2011). Prenatally, research also suggests poor maternal diet as a risk factor for child mental health problems (Jacka et al. 2013; Steenweg-de Graaff et al. 2014). In this section below, some research findings will be presented to highlight the link between specific nutritional deficiencies and different mental health outcomes.

5.4 Micronutrients and Mental Health

Given the fact that brain functions require a high metabolic rate, a relatively big proportion of a diet rich in macro- and micronutrients is dedicated to brain functions (Logan and Jacka 2014). Generally, different nutrients and polyphenols act as antioxidants, and research supports the effect of different nutrients on the mental status of individuals through their effects on structural and functional brain changes (Bourre 2006).

Vitamin B is one micronutrient contributing to mental health with research suggesting an association between vitamin B12 and folate deficiency and depressive symptoms (Tiemeier et al. 2002) and suggesting the role of vitamin B complex supplementation in enhancing both mood symptoms and the mental health quality of life in patients with depression (Lewis et al. 2013). More specifically, vitamin B12 and folic acid have also proven to reduce the risk of onset of depression in relation to aging through their impact on vascular and other metabolic risk factors (Reynolds 2002) supporting the protective role of folate and the metabolically related B-vitamins against depression in aging adults (Moore et al. 2018). Another study showed that low intake of vitamins B1, B2, B3, B5, B6, and folate and low intake of vitamin B6 and folate were associated with higher externalizing and internalizing behavior among adolescents, respectively (Herbison et al. 2012). In menstruating women, vitamin B6 has also shown to contribute to positive outcomes in relation to serotonin during the mid-luteal phase and to, therefore, help with the treatment of premenstrual syndrome (Whelan et al. 2009).

A systematic review and meta-analysis by Young et al. (2019) of 18 articles covering 16 clinical trials involving 2015 participants supported the use of vitamin B over placebo for depression in clinical populations and for stress in at-risk populations.

Food supplementation using micronutrient formulas or nutraceuticals has been implicated in mental health care as mono therapy (Rucklidge and Kaplan 2013), but more commonly as adjunctive besides drugs (Sarris et al. 2010). Micronutrient supplementation has been shown to reduce violent behavior among incarcerated juveniles with mental health problems (Schoenthaler et al. 1997). Additionally, enriching the Mediterranean diet with tryptophan and magnesium has also proved to be effective in reducing trait anxiety and eating disorders and improving self-image perception and mood among women diagnosed with fibromyalgia (Martínez-Rodríguez et al. 2020). An association between iron deficiency anemia and depression has also been demonstrated suggesting the significance of micronutrients such as iron and zinc specifically in improving mental health outcomes (Hidese et al. 2018).

Although such findings may seem promising, it is important to appreciate other research with less significant results. For example, a meta-analysis on the effects of vitamin and mineral supplementation suggested their effects on reducing levels of perceived stress, mild psychiatric symptoms, fatigue, confusion, and anxiety, but not depression (Long and Benton 2013).

5.5 PUFAs and Mental Health

Although no causal relationships have been demonstrated, some mechanisms have been proposed to explain the link between PUFAs and risk of depression. For example, depressive disorders have been linked with specific metabolism of fatty acids which consequently affect the expression of PUFA levels (Mocking et al. 2018). It has also been speculated that PUFAs are involved in the dysregulation of biological stress systems that potentially underlie depressive disorders (Thesing et al. 2018). For example, PUFAs regulate different neurotransmitter systems, regulate membrane fluidity, and promote neurite growth (Parletta et al. 2013). Intake of n-3 PUFAs is also hypothesized to influence mood by blocking the NF-κB pathway (Zhao et al. 2004). Another specific example is that of eicosapentaenoic acid (EPA) and docosahexaenoic acid (DHA) which have neuroprotective functions against cortisol-induced reduction in hippocampal neuroproliferation, neurogenesis, and apoptosis through regulating the oxidative stress and immune pathways and other pathways involved in cell development and neuronal formation (Borsini et al. 2020).

Besides reducing depressive symptoms as demonstrated by randomized controlled studies (Lespérance et al. 2011; Liao et al. 2019; Luo et al. 2020; Zhang et al. 2020b), PUFAs also play a role in modifying brain functional connectivity as suggested by findings showing a negative association between functional connectivity of the right middle frontal cortex and the right middle temporal pole and omega-3 intakes (Park et al. 2020). Interestingly, it has also been shown that an omega-3 fatty acid, DHA, is able to correct the genomic and network changes elicited by fructose in an animal model (Meng et al. 2016).

PUFA supplementation has shown to have antidepressant effects in animal models of acute and chronic stress through the regulation of neurological processes of the hypothalamic-pituitary-adrenal (HPA) axis. More specifically, it resulted in a reduction in corticotropin-releasing factor, adrenocorticotropic hormone, corticosterone, hippocampal expressions of TNF-α, and IL-6, and an increase in hippocampal glucocorticoid receptor expressions and serotonin concentrations (Hong 2020). Studies on the effectiveness of using PUFA in the management of mental health problems are controversial. Meta-analyses support the benefits of using PUFAs in psychiatric conditions, as adjuvant rather than mono therapy (Grosso et al. 2014). However, the take-home message is that, assuming their potential therapeutic use, doses should still be adjusted according to the patient's need. One study showed that omega-6/omega-3 ratio above 10:1 could predict depressive symptoms in pregnant women (de Sousa and dos Santos 2020).

5.6 Vitamin D and Mental Health

Vitamin D has proven not only to improve mental health status but also decrease inflammatory markers (Jamilian et al. 2019). One explanation is that calcitriol, as the active form of vitamin D, activates the gene expression of an enzyme involved in the synthesis of catecholamines. This indirect involvement in the synthesis of catecholamines may explain the neurotransmitter basis for the pathology of some psychiatric disorders including depression (Eyles et al. 2013).

Research also suggests the role of low maternal vitamin D during gestation in increasing the risk of mental disorders in later life (Lisi et al. 2020). Additionally, low maternal serum vitamin D during pregnancy has been studied as a risk factor for postpartum depression symptoms (Robinson et al. 2014). However, randomized controlled trials have shown inconsistent findings in relation to the use of vitamin D supplementation for mental disorders suggesting no statistical significance in reducing incidence or recurrence of depression (Okereke et al. 2020; Parker et al. 2017). In combination with the administration of SSRI, one randomized controlled trial demonstrated the effectiveness of vitamin D supplementation (Khoraminya et al. 2013).

One meta-analysis of both observational studies and randomized controlled trials with a total of 31,424 participants suggested that low vitamin D levels could predict depression with an increased odds and hazard ratios of depression in lower vitamin D level categories (Anglin et al. 2013). This is supported by another meta-analysis of 25 trials involving 7534 participants supporting the use of vitamin D supplementation (Cheng et al. 2020). Another meta-analysis of four clinical trials involving 948 participants favored the use of vitamin D supplementation, yet with moderate effect size, 0.58 (Vellekkatt and Menon 2019). On the other hand, another meta-analysis of nine clinical trials involving a total of 4923 participants did not suggest any significant effects of vitamin D supplementation on improving depressive symptoms (Gowda et al. 2015).

In general, research suggests the association between vitamin D deficiency and depression, yet there is high need for randomized controlled trials assessing nutritional interventions to examine the potential of causal mechanisms between both and to either support or reject the endorsement of vitamin D supplementation (Anglin et al. 2013).

5.7 Probiotics and Mental Health

Probiotics, referred to as psychobiotics for their effects on psychological processes (Sarkar et al. 2016), are hypothesized to be among the many mediators between nutrition and mental health. This is supported by findings showing altered microbiota in patients with major depressive disorder (Jiang et al. 2015).

Probiotics have shown to be able to regulate the activity of the HPA in response to stress (Sudo et al. 2004), decrease intestinal permeability to prevent endotoxins from entering the circulation (Ait-Belgnaoui et al. 2014), and reduce inflammatory responses to stress (Desbonnet et al. 2010).

One meta-analysis reported gut microbiome changes in individuals with serious mental disorders compared to healthy controls; however, strong empirical evidence supporting the therapeutic use of probiotics in such cases is not available (Nguyen et al. 2021).

One model explaining the mechanisms of action by which probiotic can improve gut dysbiosis and, therefore, reduce stress proposes its role in stimulating the vagus nerve, modulating immunological pathways, regulating the HPA axis which is involved in the pathology of various psychiatric conditions, down modulating oxidative stress, enhancing neurotrophic growth factors, and increasing monoamine levels (Tyagi et al. 2020). This is believed to be mediated by the positive effect of probiotics on cortisol (Takada et al. 2016). Other mechanisms of action have been proposed including the regulation of serum corticotropin-releasing factor (Yang et al. 2016) and tumor necrosis factor-α (Marcos et al. 2004).

One randomized controlled trial on expecting mothers and breastfeeding mothers 6 months after delivery showed the potential of probiotic (*Lactobacillus rhamnosus* HN001) supplementation in lowering postnatal symptoms of depression and anxiety (Slykerman et al. 2017). Similar results were demonstrated in a randomized controlled trial for the treatment of anxiety (Eskandarzadeh et al. 2021) and another one demonstrating the effect of probiotics on reactivity to sad mood (Steenbergen et al. 2015). Probiotics have also shown to reduce the overall subjective stress and to improve stress-related subthreshold anxiety/depression level, yet without having significant effects on cortisol level (Zhang et al. 2020a). On the other hand, other studies did not support such findings (See Chahwan et al. 2019) which could be attributed to the different strains, treatment schedules, samples sizes, etc. For example, one review of randomized clinical trials on the use of probiotics on depressive symptoms among adults showed inconsistent findings on the benefits of probiotics and less conclusive findings on *prebiotics*, fibers that feed probiotics, and *synbiotics*, combinations of probiotics and prebiotics (Vaghef-Mehrabany et al. 2020).

Meta-analyses on the effectiveness of probiotics have demonstrated mixed results. For example, one meta-analysis showed the effects of probiotic supplementation on decreasing inflammatory biomarkers and depressive symptoms (Amirani et al. 2020). Similar results in other meta-analyses showed larger effect sizes in clinical, rather than nonclinical populations, with therapeutic effects of probiotics but not prebiotics (Liu et al. 2019), and as adjuvant therapies rather than stand-alone treatments (Nikolova et al. 2019). On the other hand, insignificant effects were noted in other meta-analyses, probably due to methodological factors (Ng et al. 2018). Preclinical findings from animal models showed the anxiolytic effects of probiotics as opposed to clinical research in humans which did not support the effectiveness of using probiotics in patients with anxiety (Reis et al. 2018).

5.8 The Mediterranean Diet as an Example

Overall, research supports a diet rich in fruits, vegetables, nuts, fish, and olive oil, and less in meats, bakery, sugars, and trans fats for reducing the risk of depression (Huang et al. 2019), and potentially other mental health problems. One specific dietary style with such nutritional contents that has attracted research attention is the Mediterranean diet.

The CHIANTI prospective population-based study in Italy demonstrated the effects of Mediterranean diet on mood among the elderly (Milaneschi et al. 2011). Results showed higher increase of IL-6, as an inflammatory biomarker, among patients who did not adhere to a Mediterranean diet compared to those who did and, additionally, this increase was associated with a significant increase in depressive symptoms. Similar results were found in another study with a follow-up of 7.2 years which showed less occurrence of new depressive symptoms in individuals with greater adherence to a Mediterranean-based diet (Skarupski et al. 2013). Randomized controlled trials also suggest the effectiveness of Mediterranean diet in alleviating depressive symptoms (Parletta et al. 2019). More impressively, one randomized controlled trial assessing the impact of a 10-day nutrient-dense Mediterranean diet showed significant improvements in contentment (McMillan et al. 2011) and self-reported mood (Lee et al. 2015) in those who adhered to the diet change. One systemic review of twenty observational studies and six intervention trials shows that 85% of studies support the use of Mediterranean diet for the treatment of depression with high cost-effectiveness (Altun et al. 2019). Additionally, a link between better sleep quality and high adherence to the Mediterranean diet has also been observed and is potentially mediated by body weight (Godos et al. 2019). Similarly, the idea of mediation of "body weight" or, more specifically, "body image" can also explain the link between diet and mood.

The effects of the Mediterranean diet can be explained in terms of its antioxidant effect and role in decreasing pro-inflammatory cytokines, or the accumulation of visceral fat which, in itself, is capable of triggering inflammatory processes (Milaneschi et al. 2011). Therapeutically, antioxidants have proven to protect from the effects of oxidative stress underlying the pathology associated with different psychiatric disorders (Gandhi and Abramov 2012). Mediterranean-like diets have also been associated with reduced pro-inflammatory markers (Watzl et al. 2005). Another brain-based model explaining the mechanism of action by which such diets improve brain functioning illustrates the role of different nutrients (i.e., citations A,

B, C, D, and E, amino acids, oleic acid, polyphenols, omega-3 PUFA, and minerals) in increasing neuronal survival, energy metabolism, neurotransmission, membrane fluidity, etc. (Parletta et al. 2013). For example, polyphenols have proven to have immunomodulatory and anti-inflammatory effects that might mediate this relationship (Yahfoufi et al. 2018). Among the many health benefits of Mediterranean diet, it is also able to reduce levels of serum cholesterol (Wardle et al. 2000) which has been shown to improve psychological states (Wardle et al. 2000; Wells et al. 1998).

5.9 Microbiome and Mental Health

Different factors contribute to the formation of the microbiome including not only diet but also environmental and antibiotic exposures that either directly or indirectly affect mental health through the hypothalamic-adrenal-pituitary axis. And although research in the area of microbiome manipulation is promising, many methodological challenges should still be highlighted such as diagnostic considerations, the absence of one-size-fits-all probiotics, and diets as a major confounding variable in any study (Deans 2017).

Recent research has shed light on the role of gut microbiome in regulating brain functions and such implications on mental health. For instance, patients with major depression disorder (MDD) may have different fecal microbiome compared to healthy individuals including higher levels of Bacteroidetes, Protobacteria, and Actinobacteria, and less Firmicutes. Levels of genus such as less Faecalibacterium can also inversely predict the severity of depression as suggested by research (Jiang et al. 2015). Even in early life, research suggests a link between gut microbiome and temperaments in 6-month-olds (Aatsinki et al. 2019) and toddlers (Christian et al. 2015) dependent on sex. Such findings may challenge the idea that temperament is inherently inborn and may pave the way for early life interventions targeting the microbiome (Christian 2019).

5.10 Potential Pathways and Mechanisms of Action

Diet can affect mental health through both direct and indirect mechanisms. Generally, this relationship can be explained in terms of the effects of foods on glycemia, the gut microbiome, and immune activation through various hormonal, neural, and inflammatory pathways (Firth et al. 2020).

A considerable amount of attention has been given to the gut-brain-microbiota axis as a model conceptualizing the pathology of affective and stress-related disorders in terms of microbial imbalance or dysbiosis (Dinan et al. 2019). This is supported by research demonstrating the role of human gut microbiome of individuals with depression in inducing behavioral and physiological depressive symptoms in animal models (Kelly et al. 2016). This is in addition to the potential of using gut microbiota in epigenetic modification (Miro-Blanch and Yanes 2019) and altering it for therapeutic purposes through dietary changes (David et al. 2014). For instance, in one study an animal model was utilized to analyze the gut microbiome composition after the consumption of diets similar to Western diets or Mediterranean diets.

In terms of the mechanistic link between gut microbiota and mood shift, research suggests the ability of gut microbiota to affect different brain functions. For example, high production of actetate, as a product of gut microbiota, has proven to activate the parasympathetic nervous system (Perry et al. 2016). It has also been suggested that the gut microbiota can influence stress response given that the locus of gut control lies in the limbic system (Jones et al. 2006).

Other pathways of communication between the gut and brain involves the autonomic nervous system, the nerves of the gastrointestinal tract, and hypothalamic-pituitary-adrenal axis (Appleton 2018).

Animals feeding on the Mediterranean diet had more significant microbiome diversity compared to those feeding on Western diets and had higher abundance of genera Lactobacillus, Clostridium, Faecalibacterium, and Oscillospira and lower abundance of Ruminococcus and Coprococcus. On the other hand, animals feeding on the Western diet had a higher Firmicutes-Bacteroides ratio and a significantly higher abundance of families Clostridiaceae and Lactobacillaceae (Nagpal et al. 2018). Microbiota can also affect different metabolic functions (Sonnenburg & Backhed, 2016) and immune functions (Hyland & Cryan, 2016) which may explain the effect of poor diets on the development of diseases in healthy individuals.

Mental health problems can also be conceptualized from an immune perspective. Besides adiposity, one major pathway between diet and mental health is the inflammatory pathway (Oddy et al. 2018). This is confirmed by the finding that individuals with depression may display unregulated inflammatory processes and inflammatory markers such as IL-6, CRP, tumor necrosis factor alpha, and leptin (Howren and Lamkin 2009; Raison et al. 2006). Different immune cells and inflammatory molecules have also shown to affect the process of neurotransmission, neuronal death (Anisman et al. 2018), neuroendocrine function, and synaptic plasticity (Dantzer et al. 2008). Such findings may explain the molecular basis underlying the assumption that chronic inflammation may mediate the link between diet and depression (Lucas et al. 2014). Lifestyle factors such as smoking and sleep problems can also increase inflammation (Berk et al. 2013). Another mediator is stress which can significantly influence food intake and is potentially associated with pro-inflammatory processes.

In terms of metabolism, frequent and sudden increases and drops in blood glucose levels are also hypothesized to contribute to changes in psychological states (Salari-Moghaddam et al. 2019).

Moving to genetics and given individual differences in metabolism, nutrigenomics-based research is rapidly evolving to examine the nutrient-host interaction and provide a better understanding of such interactions in cases of pathology and for developing potentially effective therapeutic interventions. New findings on the genetics of food consumption and metabolism have been demonstrated from different models, and it is now established that bacterial genomics can be used, for

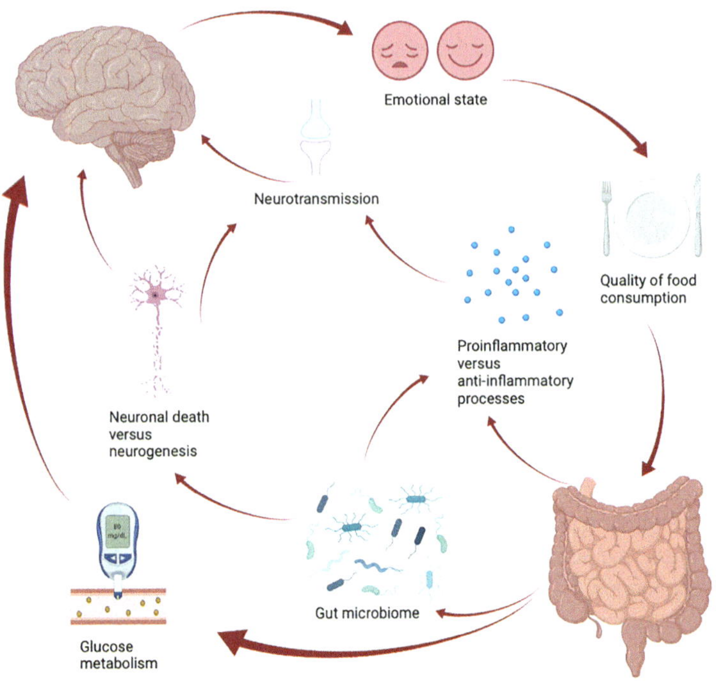

Fig. 5.1 Some mechanisms underlying the bi-dictional nature of nutrition and mental health. (Created with BioRender.com)

example, to understand the link between food fermentation and gut metabolism (Vergères 2013). Knowledge of the interactions between different nutrients and genes is required for a better understanding of the nutrigenomics of mental disorders through different metabolic pathways and homeostatic control. Nutrigenomics could be the answer to the question of why individuals respond differently to dietary changes, which is an essential step in the construction of a personalized approach integrating both genetic and epigenetic variables using omics technologies.

One rodent model of fructose consumption involving the transcriptome and epigenome sequencing of the hypothalamus and hippocampus revealed reprogramming of DNA methylation, transcript abundance, alternative splicing, and gene networks governing cell metabolism, cell communication, inflammation, and neuronal signaling as underlying the nutrient-host interaction in the pathogenesis of some brain disorders. Genes *Bgn* and *Fmod* have also shown to be involved in spatial learning and memory as two key functions affected in many brain disorders (Meng et al. 2016). Research also suggests the potential of micronutrients, such as folate, polyphenols, and vitamin B12, in the regulation of gene expression through the modification of DNA methylation given their chemical nature as methyl donors. Such findings may support a nutrient-host interaction hypothesis of mental disorders; however, more research is warranted to better understand other epigenetic processes (Fig. 5.1).

5.11 Evidence for Nutritional Interventions for Mental Health

In 2015, the International Society for Nutritional Psychiatry Research issued a statement on nutritional medicine in modern psychiatry in response to the growing research on the link between nutritional quality and mental health proposing a movement towards adopting nutritional psychiatry in integrated therapeutic paradigms as mainstream (Sarris et al. 2015). It also proposed the use of "evidence-based nutritional change" for improving mental health outcomes besides the need for rigorous research and public education focused on raising awareness (Table 5.1).

5.12 Conclusion

Depression is a global public health concern affecting the lives of millions of individuals worldwide. Diet and nutrition is a potential cause or contributor but also an avenue for intervention. This chapter provides a review of evidence suggesting that nutrition may contribute to the pathogenesis of mental disorders, including depression, and also may have a role in treatment.

Reviewing the current state of the art and despite all research findings in the field of nutritional mental health, evidence supporting the link between diet and mental health is strong. However, a research gap still exists in the absence of causal relationships between diet and mental health and with limited knowledge about the therapeutic effects of "specific dietary components." Studies on the effectiveness of diet-based intervention may prove to help bridge this gap; however, many methodological limitations have yet to be addressed including small sample sizes, different population characteristics, lack of biomarkers for participant stratification, and lack of blinding to avoid bias. To better understand how food affects mental health, well-established research needs to provide a better understanding of the mechanisms of action underlying this link including metabolism, gut-brain signaling, metabolites in the blood and other organs, cellular responses to nutrients, genetic factors mediating between nutrition and mental health, and, finally, the potential impacts of diet on the expression of different genes (Adan et al. 2019).

The use of dietary interventions for improving mental health outcomes is promising; however, empirical evidence may still be insufficient for the establishment of effective interventions and more research assessing the efficacy of such interventions is required. Findings are still limited making it inappropriate to generalize to different patient groups.

Future research is warranted to better examine the effectiveness of diet-based interventions for psychiatric conditions and to understand the long-term effects of nutrition on different aspects of mental health.

Table 5.1 The levels of evidence of studies supporting the effects of nutrition on mental health outcomes cited in this chapter

	Mediterranean diet	Micronutrients and vitamin D	PUFAs	Psychobiotics
Level 1 Randomized controlled trials Systematic reviews of randomized controlled trials Meta-analyses of randomized controlled trials	(Martínez-Rodríguez et al. 2020): Participants following a tryptophan and magnesium-enriched Mediterranean diet scored significantly less on trait anxiety, mood disturbance, and eating disorder and significantly higher on self-image perception (Parletta et al. 2019): Patients following a Mediterranean diet showed higher reduction in depressive symptoms and improved mental health quality of life at 3 months and were sustained at 6 months. Improved diet and mental health were sustained at 6 months (McMillan et al. 2011): Participants who shifted to a Mediterranean diet showed significant improvements in vigor, alertness, and contentment. Similar results were not demonstrated in terms of cognitive functioning	(Lewis et al. 2013): Patients receiving vitamin B supplementation displayed significant improvements in depressive and anxiety symptoms and scored significantly higher on a mental health scale compared to placebo with modest utility (Schoenthaler et al. 1997): Correction of low blood vitamin-mineral concentrations resulted in significant improvements in rates of violence among delinquents (Martínez-Rodríguez et al. 2020): Patients with fibromyalgia following a tryptophan and magnesium-enriched Mediterranean diet had significant improvements in terms of trait anxiety, self-image, mood disturbance, and eating disorders than patients in the control group (Okereke et al. 2020): Vitamin D supplementation did not result in significant differences in terms of incidence and recurrence of depressive	(Liao et al. 2019): This meta-analysis demonstrated the effect of omega-3 PUFAs on improving depressive symptoms. Therapeutic clinical benefits of EPA dosage of $\leq$1 g/d were shown, unlike DHA-pure and DHA-major formulations (Luo et al. 2020): This meta-analysis showed the efficacy of high-dose PUFAs to low-dose PUFAs to the placebo in the early treatment of major depressive disorder (Lespérance et al. 2011): Omega-3 supplementation resulted in symptomatic improvements in patients with major depression without comorbid anxiety disorders and was superior to placebo (Zhang et al. 2020b): Omega-3 supplementation significantly improved symptoms of depression among pregnant and postpartum women and was well tolerated Higher EPA proportion in omega-3 showed higher efficacy	(Slykerman et al. 2017): Postnatal women receiving *lactobacillus rhamnosus* HN001 probiotic treatment had significantly less depression and anxiety compared to women in the control group (Eskandarzadeh et al. 2021): Eight weeks of treatment using sertraline combined with probiotics showed superior effects to sertraline alone in decreasing symptoms of anxiety in patients with generalized anxiety disorder. No effects were observed on the quality of life (Zhang et al. 2020a): Although the effects of probiotic treatment on cortisol levels were not demonstrated, this meta-analysis shows its effect on reducing levels of stress in healthy controls and subthreshold anxiety/depression with only one study out of seven reporting side effects (Vaghef-Mehrabany et al. 2020): The effects of probiotic strains

		symptoms or mood scores compared to placebo over a median follow-up period of 5.3 years (Khoraminya et al. 2013): Vitamin D supplementation besides fluoxetine resulted in significant reductions in depressive severity compared to treatment using fluoxetine alone	than lower proportions and effects were mainly observed in cases of mild-moderate depression	on depressive symptoms were inconsistent and findings may suggest that such effects may follow a strain-specific manner
Level 2 Prospective cohort studies	(Milaneschi et al. 2011): Patients who did not adhere to a healthy diet showed higher mean increase in IL-6 levels, and depressive symptoms were associated with increased IL-6 levels only in patients not adhering to a healthy diet (Skarupski et al. 2013): Higher adherence to a Mediterranean-based diet could predict fewer newly occurring depressive symptoms over a follow-up period of 7.2 years, and the annual rate of developing depressive symptoms was 98.6% lower in people with high adherence to the diet (Lee et al. 2015): Adherence to the Mediterranean diet was associated with improved cognition, contentment, alertness, and less			

(continued)

Table 5.1 (continued)

	Mediterranean diet	Micronutrients and vitamin D	PUFAs	Psychobiotics
	confusion. Cardiovascular improvements were also observed including *a significant* decrease in augmentation pressure, but not blood flow velocity through the carotid artery			
Level 3 Case-control studies Retrospective cohort studies Systematic reviews Pilot studies	(Huang et al. 2019) (Altun et al. 2019) (Godos et al. 2019): Each point increase in the Mediterranean diet score predicted 10% higher chance of having adequate sleep quality, but only among normal and overweight participants	(Bourre 2006) (Reynolds 2002) (Moore et al. 2018) (Rucklidge and Kaplan 2013) (Parker et al. 2017) (Huang et al. 2019)	(Moore et al. 2018)	

References

Aatsinki A, Lahti L, Uusitupa L, Munukka E, Keskitalo A, Nolvi S, O'Mahony S, Pietilä S, Elo L, Eerola E, Karlsson H, Karlsson L (2019) Gut microbiota composition is associated with temperament traits in infants. Brain Behav Immun 80:849–858

Adan RAH, van der Beek EM, Buitelaar JK, Cryan JF, Hebebrand J, Higgs S, Schellekens H, Dickson SL (2019) Nutritional psychiatry: towards improving mental health by what you eat. Eur Neuropsychopharmacol 29(12):1321–1332. https://doi.org/10.1016/j.euroneuro.2019.10.011

Ait-Belgnaoui A, Colom A, Braniste V, Ramalho L, Marrot A, Cartier C, Houdeau E, Theodorou V, Tompkins T (2014) Probiotic gut effect prevents the chronic psychological stress-induced brain activity abnormality in mice. Neurogastroenterol and Motility 26(4):510–520. https://doi.org/10.1111/nmo.12295

Alam A, Voronovich Z, Carley JA (2013) A review of therapeutic uses of mirtazapine in psychiatric and medical conditions. The primary care companion for CNS disorders 15(5):PCC.13r01525. https://doi.org/10.4088/PCC.13r01525

Altun A, Brown H, Szoeke C, Goodwill AM (2019) The Mediterranean dietary pattern and depression risk: a systematic review. Neurol Psychiatry Brain Res 33:1–10. https://doi.org/10.1016/j.npbr.2019.05.007

Amirani E, Milajerdi A, Mirzaei H, Jamilian H, Mansournia MA, Hallajzadeh J, Ghaderi A (2020) The effects of probiotic supplementation on mental health, biomarkers of inflammation and oxidative stress in patients with psychiatric disorders: a systematic review and meta-analysis of randomized controlled trials. Complement Ther Med 49:102361. https://doi.org/10.1016/j.ctim.2020.102361

Anglin R, Samaan Z, Walter S, McDonald S (2013) Vitamin D deficiency and depression in adults: systematic review and meta-analysis. Br J Psychiatry 202(2):100–107. https://doi.org/10.1192/bjp.bp.111.106666

Anisman H, Hymie S, Kusnecov A (2018) The immune system and mental health. Elsevier Science & Technology, Amsterdam. ProQuest Ebook Central

Appleton J (2018) The gut-brain axis: influence of microbiota on mood and mental health. Integrative medicine (Encinitas, Calif) 17(4):28–32

Berk M, Williams LJ, Jacka FN et al (2013) So depression is an inflammatory disease, but where does the inflammation come from? BMC Med 11:200

Borsini A, Stangl D, Jeffries AR et al (2020) The role of omega-3 fatty acids in preventing glucocorticoid-induced reduction in human hippocampal neurogenesis and increase in apoptosis. Transl Psychiatry 10:219. https://doi.org/10.1038/s41398-020-00908-0

Bourre J-M (2006) Effects of nutrients (in food) on the structure and function of the nervous system: update on dietary requirements for brain, part 1: micronutrients. J Nutr Health Aging 10(5):377–385

Bremner JD, Moazzami K, Wittbrodt MT, Nye JA, Lima BB, Gillespie CF, Rapaport MH, Pearce BD, Shah AJ, Vaccarino V (2020) Diet, stress and mental health. Nutrients 12(8):2428

Chahwan B, Kwan S, Isik A, van Hemert S, Burke C, Roberts L (2019) Gut feelings: a randomised, triple-blind, placebo-controlled trial of probiotics for depressive symptoms. J Affect Disord 253:317–326. https://doi.org/10.1016/j.jad.2019.04.097. Epub 2019 May 9. PMID: 31078831

Cheng Y-C, Huang Y-C, Huang W-L (2020) The effect of vitamin D supplement on negative emotions: a systematic review and meta-analysis. Depress Anxiety 37(6):549–564. https://doi.org/10.1002/da.23025

Christian LM (2019) The gut microbiome and mental health: taking baby steps. Brain Behav Immun 81:22–23. https://doi.org/10.1016/j.bbi.2019.08.001

Christian LM, Galley JD, Hade EM, Schoppe-Sullivan S, Dush CK, Bailey MT (2015) Gut microbiome composition is associated with temperament during early childhood. Brain Behav Immun 45:118–127

Crichton GE, Elias MF, Robbins MA (2016) Association between depressive symptoms, use of antidepressant medication and the metabolic syndrome: the Maine-Syracuse study. BMC Public Health 16:502

Dantzer R, O'Connor JC, Freund GG, Johnson RW, Kelley KW (2008) From inflammation to sickness and depression: when the immune system subjugates the brain. Nat Rev Neurosci 9: 46–56

David LA, Maurice CF, Carmody RN, Gootenberg DB, Button JE, Wolfe BE et al (2014) Diet rapidly and reproducibly alters the human gut microbiome. Nature 505(7484):559–563. https://doi.org/10.1038/nature12820

Deans E (2017) Microbiome and mental health in the modern environment. J Physiol Anthropol 36: 1. https://doi.org/10.1186/s40101-016-0101-y

Desbonnet L, Garrett L, Clarke G, Kiely B, Cryan JF, Dinan TG (2010) Effects of the probiotic Bifidobacterium infantis in the ma- ternal separation model of depression. Neuroscience 170(4): 1179–1188. https://doi.org/10.1016/j.neuroscience.2010.08.005

Dinan TG, Quigley EM (2011) Probiotics in the treatment of depression: science or science fiction? Aust N Z J Psychiatry 45:1023–1025

Dinan TG, Stanton C, Long-Smith C, Kennedy P, Cryan JF, Cowan CSM, Cenit MC, van der Kamp J-W, Sanz Y (2019) Feeding melancholic microbes: MyNewGut recommendations on diet and mood. Clin Nutr 38(5):1995–2001. https://doi.org/10.1016/j.clnu.2018.11.010

Eskandarzadeh S, Effatpanah M, Khosravi-Darani K, Askari R, Hosseini AF, Reisian M et al (2021) Efficacy of a multispecies probiotic as adjunctive therapy in generalized anxiety disorder: a double blind, randomized, placebo-controlled trial. Nutr Neurosci:1–7. https://doi.org/10.1080/1028415x.2019.1598669

Eyles DW, Burne TH, J., & McGrath, J. J. (2013) Vitamin D, effects on brain development, adult brain function and the links between low levels of vitamin D and neuropsychiatric disease. Front Neuroendocrinol 34:47–64

Firth J, Gangwisch JE, Borsini A, Wootton RE, Mayer EA (2020) Food and mood: how do diet and nutrition affect mental wellbeing? BMJ 369:m2382. https://doi.org/10.1136/bmj.m2382

Gandhi S, Abramov AY (2012) Mechanism of oxidative stress in neurodegeneration. Oxidative med cell longevity. Article ID 428010:11

García-Toro M, Vicens-Pons E, Gili M, Roca M, Serrano-Ripoll MJ, Vives M, Leiva A, Yáñez AM, Bennasar-Veny M, Oliván-Blázquez B (2016) Obesity, metabolic syndrome and Mediterranean diet: impact on depression outcome. J Affect Disord 194:105–108. https://doi.org/10.1016/j.jad.2015.12.064

Godos J, Ferri R, Caraci F, Cosentino FII, Castellano S, Galvano F, Grosso G (2019) Adherence to the Mediterranean diet is associated with better sleep quality in Italian adults. Nutrients 11(5): 976

van Gool CH, Kempen GI, Penninx BW, Deeg DJ, Beekman AT, van Eijk JT (2003) Relationship between changes in depressive symptoms and unhealthy lifestyles in late middle aged and older persons: results from the longitudinal aging study Amsterdam. Age Ageing 32:81–87

Gowda U, Mutowo MP, Smith BJ, Wluka AE, Renzaho AMN (2015) Vitamin D supplementation to reduce depression in adults: meta-analysis of randomized controlled trials. Nutrition 31(3): 421–429. https://doi.org/10.1016/j.nut.2014.06.017

Grosso G et al (2014) Role of Omega-3 fatty acids in the treatment of depressive disorders: a comprehensive meta-analysis of randomized clinical trials. PLoS One 9(5) Gale In Context: Opposing Viewpoints. link.gale.com/apps/doc/A418708287/OVIC?u=aucairo&sid=summon&xid=c511c7b2. Accessed 9 Nov 2021

Herbison CE, Hickling S, Allen KL, O'Sullivan TA, Robinson M, Bremner AP, Huang R-C, Beilin LJ, Mori TA, Oddy WH (2012) Low intake of B-vitamins is associated with poor adolescent mental health and behaviour. Prev Med 55(6):634–638. https://doi.org/10.1016/j.ypmed.2012.09.014

Hidese S, Saito K, Asano S, Kunugi H (2018) Association between iron-deficiency anemia and depression: a web-based Japanese investigation. Psychiatry Clin Neurosci 72(7):513–521

Hong, J. (2020). Antidepressant effects of omega-3 PUFA on depression induced by two-hit stress in adult male offspring through hypothalamic-pituitary-adrenal axis-related mechanism. Thesis, Hanyang University. Retrieved from https://repository.hanyang.ac.kr/handle/20.500.11754/123480

Howren MB, Lamkin DM (2009) Associations of depression with C-reactive protein, IL-1, and IL-6: a meta-analysis. Suls J Psychosom Med 71:171–186

Huang Q, Liu H, Suzuki K, Ma S, Liu C (2019) Linking what we eat to our mood: a review of diet, dietary antioxidants, and depression. Antioxidants 8(9):376. https://doi.org/10.3390/antiox8090376

Hyland NP, Cryan JF (2016) Microbe-host interactions: influence of the gut microbiota on the enteric nervous system. Dev Biol. https://doi.org/10.1016/j.ydbio.2016.06.027

Jacka FN, Kremer PJ, Berk M, de Silva-Sanigorski AM, Moodie M, Leslie ER et al (2011) A prospective study of diet quality and mental health in adolescents. PLoS One 6(9):e24805. https://doi.org/10.1371/journal.pone.0024805

Jacka FN, Ystrom E, Brantsaeter AL et al (2013) Maternal and early postnatal nutrition and mental health of offspring by age 5 years: a prospective cohort study. J Am Acad Child Adolesc Psychiatry 52:1038–1047

Jamilian H, Amirani E, Milajerdi A, Kolahdooz F, Mirzaei H, Zaroudi M, Ghaderi A, Asemi Z (2019) The effects of vitamin D supplementation on mental health, and biomarkers of inflammation and oxidative stress in patients with psychiatric disorders: a systematic review and meta-analysis of randomized controlled trials. Prog Neuro-Psychopharmacol Biol Psychiatry 94: 109651. https://doi.org/10.1016/j.pnpbp.2019.109651

Jiang H, Ling Z, Zhang Y, Mao H, Ma Z, Yin Y et al (2015) Altered fecal microbiota composition in patients with major depressive disorder. Brain Behav Immun 48:186–194

Jones MP, Dilley JB, Drossman D, Crowell MD (2006) Brain-gut connections in functional GI disorders: anatomic and physiologic relationships. Neurogastroenterol Motil 18(2):91–103. https://doi.org/10.1111/j.1365-2982.2005.00730

Kachani AT, de Lima Furtado YA (2020) Diet, nutrition, and women's mental health. In: Rennó J Jr, Valadares G, Cantilino A, Mendes-Ribeiro J, Rocha R, Geraldo da Silva A (eds) Women's mental health. Springer, Cham. https://doi.org/10.1007/978-3-030-29081-8_19

Kelly JR, Borre Y, O'Brien C, Patterson E, El Aidy S, Deane J, Kennedy PJ, Beers S, Scott K, Moloney G, Hoban AE, Scott L, Fitzgerald P, Ross P, Stanton C, Clarke G, Cryan JF, Dinan TG (2016) Transferring the blues: depression-associated gut microbiota induces neurobehavioural changes in the rat. J Psychiatr Res 82:109–118. https://doi.org/10.1016/j.jpsychires.2016.07.019

Khoraminya N, Tehrani-Doost M, Jazayeri S, Hosseini A, Djazayery A (2013) Therapeutic effects of vitamin D as adjunctive therapy to fluoxetine in patients with major depressive disorder. Aust N Z J Psychiatry 47:271–275

Lee J, Pase M, Pipingas A, Raubenheimer J, Thurgood M, Villalon L, Macpherson H, Gibbs A, Scholey A (2015) Switching to a 10-day Mediterranean-style diet improves mood and cardiovascular function in a controlled crossover study. Nutrition 31(5):647–652. https://doi.org/10.1016/j.nut.2014.10.008

Lespérance F, Frasure-Smith N, St-André E, Turecki G, Lespérance P, Wisniewski SR (2011) The efficacy of omega-3 supplementation for major depression: a randomized controlled trial. J Clin Psychiatry 72(8):1054–1062. https://doi.org/10.4088/JCP.10m05966blu

Lewis JE, Tiozzo E, Melillo AB, Leonard S, Chen L, Mendez A, Woolger JM, Konefal J (2013) The effect of methylated vitamin B complex on depressive and anxiety symptoms and quality of life in adults with depression. ISRN Psychiatry. https://doi.org/10.1155/2013/621453

Liao Y, Xie B, Zhang H, He Q, Guo L, Subramaniapillai M, Fan B, Lu C, Mclntyer RS (2019) Efficacy of omega-3 PUFAs in depression: a meta-analysis. Transl Psychiatry 9(1):190. https://doi.org/10.1038/s41398-019-0515-5

Lisi G, Ribolsi M, Siracusano A, Niolu C (2020) Maternal vitamin D and its role in determining fetal origins of mental health. Curr Pharm Des 26(21):2497–2509. https://doi.org/10.2174/1381612826666200506093858

Liu RT, Walsh RFL, Sheehan AE (2019) Prebiotics and probiotics for depression and anxiety: a systematic review and meta-analysis of controlled clinical trials. Neurosci Biobehav Rev 102: 13–23. https://doi.org/10.1016/j.neubiorev.2019.03.023

Logan AC, Jacka FN (2014) Nutritional psychiatry research: an emerging discipline and its intersection with global urbanization, environmental challenges and challenges and the evolutionary mismatch. J Physiol Anthropol 33:22

Long SJ, Benton D (2013) Effects of vitamin and mineral supplementation on stress, mild psychiatric symptoms, and mood in nonclinical samples: a meta-analysis. Psychosom Med 75(2):144–153. https://doi.org/10.1097/PSY.0b013e31827d5fbd

Lopresti AL, Hood SD, Drummond PD (2013) A review of lifestyle factors that contribute to important pathways associated with major depression: diet, sleep and exercise. J Affect Disord 148(1):12–27

Lucas M, Chocano-Bedoya P, Shulze MB, Mirzaei F, O'Reilly ÉJ, Okereke OI, Hu FB, Willett WC, Ascherio A (2014) Inflammatory dietary pattern and risk of depression among women. Brain Behav Immun 36:46–53

Luo XD, Feng JS, Yang Z et al (2020) High-dose omega-3 polyunsaturated fatty acid supplementation might be more superior than low-dose for major depressive disorder in early therapy period: a network meta-analysis. BMC Psychiatry 20:248. https://doi.org/10.1186/s12888-020-02656-3

Marcos A, Wärnberg J, Nova E, Gómez S, Alvarez A, Alvarez R et al (2004) The effect of milk fermented by yogurt cultures plus lactobacillus casei DN-114001 on the immune response of subjects under academic examination stress. Eur J Nutr 43(6):381–389. https://doi.org/10.1007/s00394-004-0517-8

Martínez-Rodríguez A, Rubio-Arias JÁ, Ramos-Campo DJ, Reche-García C, Leyva-Vela B, Nadal-Nicolás Y (2020) Psychological and sleep effects of tryptophan and magnesium-enriched Mediterranean diet in women with fibromyalgia. Int J Environ Res Public Health 17(7):2227

McMillan L, Owen L, Kras M, Scholey A (2011) Behavioural effects of a 10-day Mediterranean diet. Results from a pilot study evaluating mood and cognitive performance. Appetite 56:143–147. https://doi.org/10.1016/j.appet.2010.11.149

Meng Q, Ying Z, Noble E, Zhao Y, Agrawal R, Mikhail A, Zhuang Y, Tyagi E, Zhang Q, Lee JH, Morselli M, Orozco L, Guo W, Kilts TM, Zhu J, Zhang B, Pellegrini M, Xiao X, Young MF, Gomez-Pinilla F, Yang X (2016) Systems nutrigenomics reveals brain gene networks linking metabolic and brain disorders. EBio Med 7:157–166

Milaneschi Y, Bandinelli S, Penninx BW, Vogelzangs N, Corsi AM, Lauretani F, Kisialiou A, Vazzana R, Terracciano A, Guralnik JM, Ferrucci L (2011) Depressive symptoms and inflammation increase in a prospective study of older adults: a protective effect of a healthy (Mediterranean-style) diet. Mol Psychiatry 16:589–590. https://doi.org/10.1038/mp.2010.113

Miro-Blanch J, Yanes O (2019) Epigenetic regulation at the interplay between gut microbiota and host metabolism. Front Genet 10:638

Mizuki Y, Sakamoto S, Okahisa Y, Yada Y, Hashimoto N, Takaki M, Yamada N (2021) Mechanisms underlying the comorbidity of schizophrenia and type 2 diabetes mellitus. Int J Neuropsychopharmacol 24(5):367–382. https://doi.org/10.1093/ijnp/pyaa097

Mocking RJT, Assies J, Ruhé HG, Schene AH (2018) Focus on fatty acids in the neurometabolic pathophysiology of psychiatric disorders. J Inherit Metab Dis 41:597–611. https://doi.org/10.1007/s10545-018-0158-3

Moore K, Hughes C, Ward M, Hoey L, McNulty H (2018) Diet, nutrition and the ageing brain: current evidence and new directions. Proc Nutr Soc 77(2):152–163. https://doi.org/10.1017/S0029665117004177

Nagpal Ravinder, Shively CA, Appt SA, Register TC, Michalson KT, Vitolins MZ, Hariom Y (2018) Gut microbiome composition in non-human primates consuming a Western or Mediterranean diet. Front Nutr 5:28. https://doi.org/10.3389/fnut.2018.00028

Ng QX, Peters C, Ho CYX, Lim DY, Yeo W-S (2018) A meta-analysis of the use of probiotics to alleviate depressive symptoms. J Affect Disord 228:13–19. https://doi.org/10.1016/j.jad.2017.11.063

Nguyen TT, Hathaway H, Kosciolek T, Knight R, Jeste DV (2021) Gut microbiome in serious mental illnesses: a systematic review and critical evaluation. Schizophr Res 234:24–40. https://doi.org/10.1016/j.schres.2019.08.026

Nikolova V, Zaidi SY, Young AH, Cleare AJ, Stone JM (2019) Gut feeling: randomized controlled trials of probiotics for the treatment of clinical depression: systematic review and meta-analysis. Therap Adv Psychopharmacol. https://doi.org/10.1177/2045125319859963

Oddy WH, Allen KL, Trapp GSA, Ambrosini GL, Black LJ, Huang R-C, Rzehak P, Runions KC, Pan F, Beilin LJ, Mori TA (2018) Dietary patterns, body mass index and inflammation: pathways to depression and mental health problems in adolescents. Brain Behav Immun 69: 428–439. https://doi.org/10.1016/j.bbi.2018.01.002

Okereke OI, Reynolds CF, Mischoulon D et al (2020) Effect of long-term vitamin D3 supplementation vs placebo on risk of depression or clinically relevant depressive symptoms and on change in mood scores: a randomized clinical trial. JAMA 324(5):471–480. https://doi.org/10.1001/jama.2020.10224

Park S-J, Lee D-K, Kim B, Na K-S, Lee C-H, Son Y-D, Lee H-J (2020) The association between omega-3 fatty acid intake and human brain connectivity in middle-aged depressed women. Nutrients 12(8):2191. https://doi.org/10.3390/nu12082191

Parker GB, Brotchie H, Graham RK (2017) Vitamin D and depression. J Affect Disord 208:56–61. https://doi.org/10.1016/j.jad.2016.08.082

Parletta N, Milte CM, Meyer BJ (2013) Nutritional modulation of cognitive function and mental health. J Nutr Biochem 24(5):725–743. https://doi.org/10.1016/j.jnutbio.2013.01.002

Parletta N, Zarnowiecki D, Cho J, Wilson A, Bogomolova S, Villani A, Itsiopoulos C, Niyonsenga T, Blunden S, Meyer B, Segal L, Baune BT, O'Dea K (2019) A Mediterranean-style dietary intervention supplemented with fish oil improves diet quality and mental health in people with depression: a randomized controlled trial (HELFIMED). Nutr Neurosci 22(7): 474–487. https://doi.org/10.1080/1028415X.2017.1411320

Perry RJ et al (2016) Acetate mediates a microbiome–brain–beta-cell axis to promote metabolic syndrome. Nature 534:213–217

Raison CL, Capuron L, Miller AH (2006) Cytokines sing the blues: inflammation and the pathogenesis of depression. Trends Immunol 27:24–31

Reis DJ et al (2018) The anxiolytic effect of probiotics: a systematic review and meta-analysis of the clinical and preclinical literature. PLoS One 13(6):e0199041. Gale In Context: Opposing Viewpoints, link.gale.com/apps/doc/A543720014/OVIC?u=aucairo&sid=summon&xid=33df0bd9. Accessed 9 Nov 2021

Reynolds EH (2002) Folic acid, ageing, depression, and dementia. BMJ 324:1512–1515

Robinson M, Whitehouse AJO, Newnham JP et al (2014) Low maternal serum vitamin D during pregnancy and the risk for postpartum depression symptoms. Arch Womens Ment Health 17: 213–219. https://doi.org/10.1007/s00737-014-0422-y

Rucklidge JJ, Kaplan BJ (2013) Broad-spectrum micronutrient formulas for the treatment of psychiatric symptoms: a systematic review. Expert Rev Neurother 13(1):49–73

Salari-Moghaddam A, Saneei P, Larijani B, Esmaillzadeh A (2019) Glycemic index, glycemic load, and depression: a systematic review and meta-analysis. Eur J Clin Nutr 73:356–365. https://doi.org/10.1038/s41430-018-0258-z

Sarkar A, Lehto SM, Harty S, Dinan TG, Cryan JF, Burnet PWJ (2016) Psychobiotics and the manipulation of bacteria-gut-brain signals. Trends Neurosci 39(11):763–781. https://doi.org/10.1016/j.tins.2016.09.002

Sarris J, Kavanagh DJ, Byrne G (2010) Adjuvant use of nutritional and herbal medicines with antidepressants, mood stabilizers and benzodiazepines. J Psychiatr Res 44(1):32–41

Sarris J, Logan AC, Akbaraly TN, Amminger GP, Balanzá-Martínez V, Freeman MP, Hibbeln J, Matsuoka Y, Mischoulon D, Mizoue T, Nanri A, Nishi D, Ramsey D, Rucklidge JJ, Sanchez-Villegas A, Scholey A, Su K-P, Jacka FN (2015) Nutritional medicine as mainstream in psychiatry. Lancet Psychiatry 2(3):271–274. https://doi.org/10.1016/S2215-0366(14)00051-0

Schoenthaler SJ, Amos S, Doraz W, Kelly M-A, Muedeking G, Wakefield J (1997) The effect of randomized vitamin-mineral supplementation on violent and non-violent antisocial behavior among incarcerated juveniles. J Nutr Environ Med 7:343–352

Skarupski KA, Tangney CC, Li H et al (2013) Mediterranean diet and depressive symptoms among older adults over time. J Nutr Health Aging 17:441–445. https://doi.org/10.1007/s12603-012-0437-x

Slykerman RF, Hood F, Wickens K, Thompson JMD, Barthow C, Murphy R, Kang J, Rowden J, Stone P, Crane J, Stanley T, Abels P, Purdie G, Maude R, Mitchell EA (2017) Effect of lactobacillus rhamnosus HN001 in pregnancy on postpartum symptoms of depression and anxiety: a randomised double-blind placebo-controlled trial. EBioMedicine 24:159–165. https://doi.org/10.1016/j.ebiom.2017.09.013

Sonnenburg JL, Bäckhed F (2016) Diet-microbiota interactions as moderators of human metabolism. Nature 535(7610):56–64. https://doi.org/10.1038/nature18846. PMID: 27383980; PMCID: PMC5991619

de Sousa TM, dos Santos LC (2020) Dietary fatty acids, omega-6/omega-3 ratio and cholesterol intake associated with depressive symptoms in low-risk pregnancy. Nutr Neurosci. https://doi.org/10.1080/1028415X.2020.1792618

Steenbergen L, Sellaro R, van Hemert S, Bosch JA, Colzato LS (2015) A randomized controlled trial to test the effect of multispecies probiotics on cognitive reactivity to sad mood. Brain Behav Immun. https://doi.org/10.1016/j.bbi.2015.04.003

Steenweg-de Graaff J, Tiemeier H, Steegers-Theunissen RP et al (2014) Maternal dietary patterns during pregnancy and child internalising and externalising problems. Gener R Study Clin Nutr 33:115–121

Sudo N, Chida Y, Aiba Y, Sonoda J, Oyama N, Yu X-N et al (2004) Postnatal microbial colonization programs the hypotha- lamic-pituitary-adrenal system for stress response in mice. J Physiol 558(Pt 1):263–275. https://doi.org/10.1113/jphysiol.2004.063388

Takada M, Nishida K, Kataoka-Kato A, Gondo Y, Ishikawa H, Suda K et al (2016) Probiotic lactobacillus casei strain Shirota relieves stress-associated symptoms by modulating the gut-brain interaction in human and animal models. Neurogastroenterol Motil 28(7):1027–1036. https://doi.org/10.1111/nmo.12804

Teasdale SB, Samaras K, Wade T, Jarman R, Ward PB (2017) A review of the nutritional challenges experienced by people living with severe mental illness: a role for dietitians in addressing physical health gaps. J Hum Nutr Diet 30:545–553

Thesing CS, Bot M, Milaneschi Y, Giltay EJ, Penninx BWJH (2018) Omega-3 polyunsaturated fatty acid levels and dysregulations in biological stress systems. Psychoneuroendocrinology 97:206–215. https://doi.org/10.1016/j.psyneuen.2018.07.002

Tiemeier H, van Tuijl HR, Hofman A, Meijer J, Kiliaan AJ, Breteler MMB (2002) Vitamin B12, folate, and homocysteine in depression: the Rotterdam study. Am J Psychiatr 159(12):2099–2101

Tyagi P, Tasleem M, Prakash S, Chouhan G (2020) Intermingling of gut microbiota with brain: exploring the role of probiotics in battle against depressive disorders. Food Res Int 137:109489. https://doi.org/10.1016/j.foodres.2020.109489

Vaghef-Mehrabany E, Maleki V, Behrooz M, Ranjbar F, Ebrahimi-Mameghani M (2020) Can psychobiotics "mood"ify gut? An update systematic review of randomized controlled trials in healthy and clinical subjects, on anti-depressant effects of probiotics, prebiotics, and synbiotics. Clin Nutr 39(5):1395–1410. https://doi.org/10.1016/j.clnu.2019.06.004

Vellekkatt F, Menon V (2019) Efficacy of vitamin D supplementation in major depression: a meta-analysis of randomized controlled trials. J Postgrad Med 65(2):74–80. https://doi.org/10.4103/jpgm.JPGM_571_17
Vergères G (2013) Nutrigenomics – linking food to human metabolism. Trends Food Sci Technol 31:6–12. https://doi.org/10.1016/j.tifs.2013.02.002
Wardle J, Rogers P, Judd P, Taylor MA, Rapoport L, Green M, Perry KN (2000) Randomized trial of the effects of cholesterol-lowering dietary treatment on psychological function. Am J Med 108(7):547–553. https://doi.org/10.1016/S0002-9343(00)00330-2
Watzl B, Kulling SE, Moseneder J et al (2005) A 4-wk intervention with high intake of carotenoid-rich vegetables and fruit reduces plasma C-reactive protein in healthy, nonsmoking men. Am J Clin Nutr 82:1052–1058
Wells AS, Read NW, Laugharne JD et al (1998) Alterations in mood after changing to a low-fat diet. Br J Nutr 79:23–30
Whelan AM, Jurgens TM, Naylor H (2009) Herbs, vitamins and minerals in the treatment of premenstrual syndrome: a systematic review. Can J Clin Pharmacol 16(3):e407–e429
Yahfoufi N, Alsadi N, Jambi M, Matar C (2018) The immunomodulatory and anti-inflammatory role of polyphenols. Nutrients 10:E1618. https://doi.org/10.3390/nu10111618. pmid:30400131
Yang H, Zhao X, Tang S, Huang H, Zhao X, Ning Z et al (2016) Probiotics reduce psychological stress in patients before laryngeal cancer surgery. Asia-Pacific J Clin Oncol 12(1):e92–e96. https://doi.org/10.1111/ajco.12120
Young LM, Pipingas A, White DJ, Gauci S, Scholey A (2019) A systematic review and meta-analysis of B vitamin supplementation on depressive symptoms, anxiety, and stress: effects on healthy and 'at-risk' individuals. Nutrients 11(9):2232. https://doi.org/10.3390/nu11092232
Zhang N, Zhang Y, Li M et al (2020a) Efficacy of probiotics on stress in healthy volunteers: a systematic review and meta-analysis based on randomized controlled trials. Brain Behav 10: e01699. https://doi.org/10.1002/brb3.1699
Zhang MM, Zou Y, Li SM, Wang L, Sun YH, Shi L, Lu L, Bao YP, Li SX (2020b) The efficacy and safety of omega-3 fatty acids on depressive symptoms in perinatal women: a meta-analysis of randomized placebo-controlled trials. Transl Psychiatry 10(1):193. https://doi.org/10.1038/s41398-020-00886-3
Zhao Y, Joshi-Barve S, Barve S et al (2004) Eicosapentaenoic acid prevents LPS-induced TNF-alpha expression by preventing NF-kappaB activation. J Am Coll Nutr 23:71–78

Chapter 6
Nutrigenomics and Neurodevelopmental Disorders

Sara Ayman Moustafa, Reem Deif, and Hassan El-Fawal

6.1 Introduction

Over the past decades, recent research has reignited interest not only in the genetic risk factors associated with some neurodevelopmental deficits but also in the role of nutrition and the mechanisms by which different genetic makeups may contribute to biological responses to dietary styles in neurodevelopmental disorders. Several nutrients and genetic profiles have been proposed to demonstrate such models; yet many controversies still exist. Several research findings have paved the way for research in the nutrigenomics of neurodevelopmental disorders. One model that attempts to explain the link between nutrition and gene expression is the mercury toxicity model (Dufault et al. 2009), according to which transcription factors are determined by the sufficiency or insufficiency of some nutrients or the presence of some toxic substances, which then influences whether or not a gene will be expressed. This chapter provides a brief overview of the nutrigenomics of neurodevelopmental disorders, explaining some proposed links, biological mechanisms of action, and finally providing examples of two common neurodevelopmental disorders: autism spectrum disorder (ASD) and attention-deficit hyperactivity disorder (ADHD).

S. A. Moustafa (✉) · R. Deif · H. El-Fawal
Institute of Global Health and Human Ecology, School of Sciences and Engineering, The American University of Cairo, New Cairo, Egypt
e-mail: saraayman@aucegypt.edu

M. Salama (ed.), *Nutrigenomics and the Brain*, Nutritional Neurosciences,
https://doi.org/10.1007/978-981-16-9205-5_6

Gut and vaginal microbiome are essential players during pregnancy, ensuring proper fetal development and microbiome handover at birth, which affects both maternal and fetal health. Intestinal microbiota face reduction in its diversity during pregnancy progression, where *Proteobacteria* is enriched. This enrichment is crucial to respond to the increased metabolic demands of a growing fetus. Newborn microbiota in early life, depending on different nutritional aspects, can determine the newborn's early metabolic programming. For instance, it was reported that breastfed and formula-fed infants have distinctive differences in their gut microbiota. Exclusively breastfed infants are characterized with an abundance of *Bacteroides* and *Bifidobacterium* in their gut composition compared to formula-fed ones. Another benefit of breastfeeding that has been reported in research is the enhanced verbal and nonverbal cognitive functions later during childhood (Cerdó et al. 2019; Codagnone et al. 2019).

6.5 Iron Deficiency

In a world where 25% of its population are affected by anemia, 50% of them suffer from iron-deficiency anemia. Children are the most vulnerable population to iron deficiency. Being crucial for brain metabolism and development, its deficiency can cause alteration in the neurotransmitter homeostasis, impaired synaptogenesis, and decline in basal ganglia function and myelin production (Lozoff 2011). On the cognitive and behavioral levels, psychomotor functions and mental capacity are affected by iron deficiency. Numerous studies have reported a link between iron deficiency and psychomotor development impairment in children (Wiegersma et al. 2019). Children younger than 10 years old with iron deficiency at an early age are at high risk of moderate mental retardation, and those with hemoglobin levels below 100 g/L scored lower on tests for intellectual development rates (Chmielewska et al. 2019). Based on such findings, the World Health Organization has recommended iron fortification and iron supplementation programs to counteract cognitive impairment (Villamor et al. 2012).

6.6 Ketogenic Diet

The ketogenic diet is a form of fasting aimed at producing ketones and altering energy metabolism. It is a high-fat, low-carbohydrate, and moderate-protein diet that uses fat as a fuel instead of glucose (Kraeuter et al. 2019). This has shown to reduce insulin signaling and circulating insulin resulting in a metabolic shift in the fatty acid metabolization causing an increased ketone production. Research suggests that it can have several preventative properties in relation to neurodegenerative and neurodevelopmental diseases. Limited data on the effect of ketogenic diet in neurodevelopmental disorders is available (Ly et al. 2017). Animal studies have shown that a ketogenic diet causes reduction in activity levels compared to control animals without ADHD. Limited, yet promising, results from human studies reporting the effect of ketogenic diet on ADHD and autism spectrum in children have been reported (Gogou and Kolios 2018; Kraeuter et al. 2019).

6.7 Examples of Neurodevelopmental Disorders

6.7.1 Autism

Autism spectrum disorder (ASD) has been among the most controversial disorders when it comes to etiology with long-held views about hypothesized genetic basis, on the one hand, and, on the other hand, it being the result of poor parenting or pathological upbringing. Given its indefinite etiology, ASD is among the idiopathic neurodevelopmental disorders that only has some hypothesized risk and protective factors (Cannell 2017). Most importantly, modern research has been able to acknowledge the role of both genetics and environmental exposure in the etiology of ASD and, more importantly, the interplay between both factors.

6.7.1.1 Prenatal Exposures, Valproic, Folic Acids, and Iron

One remarkable finding is that prenatal vitamins have been shown to reduce the risk of ASD in a dose-dependent manner (Schmidt et al. 2011). Prenatal exposure to valproic acid, on the other hand, has shown to increase its risks (Dean et al. 2002; Moore et al. 2000; Ornoy 2009). This is due to the well-established evidence that this risk is associated with neural tube defects (Dean et al. 2002) and that, since valproic acid antagonizes folic acid, folic acid fortification has proven to reduce such risks. Another review of literature suggests that maternal deficiency in iron, zinc, and copper can contribute to the risk of neurodevelopmental problems which might increase the risk of autism (Bölte et al. 2019). More specifically, research suggests the involvement of zinc in the gut development and morphology which may explain the prevalence of gastrointestinal problems in patients with autism spectrum disorders (Vela et al. 2015) without suggesting specific comorbid or causative relations (Buie et al. 2010; Coury et al. 2012). Another explanation is the hypothesized composition of gut microbiota. Interestingly, GI symptoms have also shown to positively correlate with severity of autistic symptoms (Adams et al. 2011).

Low iron intake during pregnancy can increase the risk of ASD development. Research suggests that mothers who gave birth to a child with ASD are less prone to take iron-fortified foods or supplements during the specified period and general have significantly lower daily iron intake (51.7 mg/day SD 34.0) than mothers who gave birth to a child without ASD (57.1 mg/day, SD 36.6). It is also suggested that iron supplementation in children with ASD may have a beneficial effect on sleep quality. The results of an 8-week clinical trial with an intake of oral iron supplement in a dose of 6 mg/kg/day showed that the occurrence of restless sleep improved significantly in 29% of children with ASD. Nonetheless, the delayed sleep onset improved insignificantly (35% of children as compared with 44% at baseline) (Dosman et al. 2007; Pivina et al. 2019).

6.7.1.2 Vitamin D Deficiency

Research findings suggest a somehow ambiguous link between vitamin D deficiency and risk of ASD. On the one hand, children with ASD are frequently reported to have GI problems which can be attributed to innate immunity and metabolism, and on the other hand, vitamin D and folate were found to be involved in GI functions including mucosal immune function, intestinal motility, and intestinal permeability. More directly, children with ASD have been reported to have lower levels of folate and vitamin D (Tan et al. 2020). Among the many impacts of gestational vitamin D deficiency is the decreased fetal volumes of cerebral gray and white matter. Various animal studies have reported similar altered brain morphology which were later supported by research involving patients with gestational vitamin D deficiency being correlated with specific brain morphology (Darling et al. 2017; D'Ambrosio et al. 2017; Dhamayanti et al. 2019; Zou et al. 2020).

Association occurs between bound VDR and retinoid X receptor to form a complex that acts as a transcription factor in the nucleus. It is worth noting that vitamin D upregulates its own receptor. To date, the known targets for VDR include 1,25(OH)2D3-regulated VDR specific, cathelicidin precursor, B-defensin, Cyp24 hydroxylase gene, and antimicrobial peptide (Gombart et al. 2005).

The link between vitamin D deficiency and ASD potentially comes from observations such as the observation of intellectual deficiencies among children with rickets (Gilmour 1938; Hallerhan 1938) and the finding that dark-skinned immigrants in the northern latitudes had higher incidence of autism (Cannell 2008). Animal models also support such findings; for example, behavioral phenotypes of ASD have been observed in vitamin D-deficient offsprings including impaired motor development and motor control, reciprocal social interaction, and later on hyperactivity (Ali et al. 2019) and suggesting structural and functional changes in the brain that are linked with autism (Eyles et al. 2009; Grecksch et al. 2009; Levenson and Figueiroa 2008).

Vitamin D is made available in the body mainly through the conversion of 7-dehydrocholetserol by ultraviolet light or through food intake like dairy and oily fish. Where it's taken up by the body as cholecalciferol (Prosser and Jones 2004) to be then hydroxylated twice into 1,25-hydroxyvitamin D (1,25(OH)2,D3) where it binds to the vitamin D receptor (VDR) found in the cytoplasm (Pojednic et al. 2015; Zhong et al. 2014). Association occurs between the bound VDR and retinoid X receptor to form a complex that acts as a transcription factor in the nucleus. It is worth noting that Vitamin D upregulates its own receptor. To date, the known targets for VDR includes,1,25(OH)2D3-regulated VDR specific, cathelicidin precursor, B-defensin, Cyp24 hydroxylase gene and antimicrobial peptide (Gombart et al.

2005). Variations of the VDR gene in humans have also shown to be associated with shaping gut microbiome and the abundance of *Parabacteroides*.

Special interest has gone to more specific mediators including the vitamin D receptor (VDR) which has proven to be involved in brain development (Eyles et al. 2013) as early as during embryonic development. Different polymorphisms affecting VDR gene expression have been identified with potential impact on ASD risk. For example, one example is a Chinese case-controlled study exploring the association between single nucleotide polymorphisms (SNPs) in genes encoding vitamin D metabolism-related enzymes and ASD in a sample of 249 children with ASD and 353 healthy controls (Yu et al. 2020). Results showed an association between the G/A genotype or the G allele of CYP24A1 rs17219315 and the G/A genotype of CYP27B1 rs4646536 and an increased risk of ASD. The A allele of both CYP2R1 rs12794714 and CYP27B1 rs4646536 was also significantly predictive of the severity of ASD. Similarly and besides child GC AA-genotype/A-allele, another case-controlled study suggested a significant association between paternal *VDR TaqI* homozygous variant genotype and ASD case control. On the other hand, a negative association has been demonstrated between ASD and child *CYP2R1* AA-genotype (Schmidt et al. 2015). One GWAS study reported that 130 transcripts that are immunity related were upregulated, while 22 genes that were linked to neurodevelopment were downregulated besides the upregulation of innate immune-like NK and CD8 + T cells. Responses were found to be upregulated where they were associated with antigen-specific immune response, cell death, autoimmune diseases, inflammation, and migration pathways (Balta et al. 2018). Researchers have also demonstrated a link between some polymorphisms of the VDR genes and severity of ASD symptoms (Schmidt et al. 2015). However, other studies have only shown minimal or preliminary evidence (Mobasheri et al. 2020) which shows the need for more extensive research in this area.

Vitamin D effect on fetal neurodevelopment can be done via various mechanisms, one of them proposing that vitamin D possess an anti-inflammatory effect on the brain through reducing neuroinflammation that is caused by toxins and oxidants and decreasing the levels of harmful inflammatory cytokines, increasing T-regulatory cells and glutathione upregulation, improving DNA repair, and increasing seizure threshold and neural mitochondrial protection. Vitamin D was reported to have a crucial role in serotonin neurotransmitter levels during development, as it regulates the limiting enzymes of genetic serotonin. On the other hand, vitamin D increases peripheral serotonin through the downregulation of peripheral tryptophan hydroxylase TPH1 and upregulating central TPH2 which also results in a decrease in central serotonin levels that can be a specific characteristic of ASD. Another neurotransmitter that is affected by vitamin D is dopamine, where vitamin D affects the rate-limiting enzyme for dopamine synthesis and tyrosine hydroxylase where it was reported to have a related polymorphism in animal models (Li et al. 2018).

Mechanisms of action explaining the link between vitamin D deficiency and ASD include the role of vitamin D in anti-inflammatory processes in the brain. For example, calcitriol does not only have anti-inflammatory processes but is also capable of enhancing DNA repair besides having anti-autoimmune effects (Huang et al. 2015). Looking into the hypothesized mechanism of action by which vitamin D may be associated with ASD, some also suggest the role of vitamin D in reducing inflammatory cytokines through its anti-inflammatory action in the brain (Huang et al. 2015). Similarly, calcitriol, a synthetic form of vitamin D, has proved to contribute to DNA repair mechanisms with some anti-autoimmune and antioxidant effects (Kočovská et al. 2012). Another interesting finding relates to the mechanism by which vitamin D has functional effects on enzymes that are involved in rate-limiting reaction in the biosynthesis of serotonin; this includes the role of activated vitamin D in downregulating peripheral TPH1 and upregulating central TPH2 which is especially evident in cases with ASD (Patrick and Ames 2014).

Other pathogenic mechanisms include the effect of vitamin D deficiency in altering steroidogenesis and fetal brain anatomy given its role in cell proliferation and differentiation, immunomodulation, and regulation of neurotransmission (Ali et al. 2018).

Therapeutically, research suggests the therapeutic value of vitamin D supplementation in reducing the severity of autistic symptoms as measured by the Autism Behavior Checklist (ABC) and the Childhood Autism Rating Scale (CARS), especially in younger patients (Feng et al. 2017). One animal model demonstrates the effect of vitamin D supplementation during pregnancy on preventing ASD-like behaviors in offsprings (Vuillermot et al. 2017). Randomized controlled trials also suggest the tolerance of large doses of vitamin D (300 IU vitamin D3/kg/day, not to exceed 5000 IU/day) by children with ASD and significant symptomatic improvements at 4-month follow-ups (Saad et al. 2018) and in alleviating ASD symptoms even when serum serotonin and interleukin-6 remained unchanged following treatment (Javadfar et al. 2020). One systematic meta-analysis supports the same findings (Song et al. 2020). However, other studies were not able to demonstrate similar findings suggesting the need for more extensive research (Kerley et al. 2017).

6.7.1.3 Ketogenic Diet

Preclinical studies on the effect of ketogenic diet in the ASD animal models have reported improvements in core symptoms including improved sociability, communication, and decreased repetitive behaviors. Ketogenic diet was found to have the ability to alter mitochondrial bioenergetics, hormonal metabolism, gut microbiome, and neurotransmitter signaling which can be attributed to improvement in behavioral symptoms. The effects of ketogenic diet were studied in children with ASD and showed mild to moderate improvement in symptoms.

6.7.1.4 Gluten-Free and Casein-Free Diet

The role of gluten and casein was highlighted following the hypothesis that the abnormal metabolism of these proteins increased opioid activity in the CNS, altering its functions. A leaky gut is another suggested mechanism. An increase in the passage of gluten and casein in the bloodstream due to the altered function of gut permeability and possibly blood-brain barrier as well can be correlated with autistic symptoms development. A gluten-casein-free diet is a potential alternative for ASD. GFCF diet showed little evidence for managing symptoms of ASD; it is still commonly used as a treatment for ASD in children. One study showed significant improvements in the communication subdomain of ASD for the social interaction subdomain of the Gilliam Autism Rating Scale. Daily living skills domain presented in the Vineland Adaptive Behavior Scale and inattention and hyperactivity subscales of the ADHD-IV scale more studies were improved based on parent-based assessments. More studies with large-scale RCTs are essential to better understand the impact of GFCF diet in ASD (Piwowarczyk et al. 2018).

6.7.1.5 Probiotics

Probiotics, defined as substances containing live microorganisms with beneficial health effects when provided in adequate amounts (FAO/WHO 2001), are among the factors that have proven to be effective in improving gut microbial, gastrointestinal, and even behavioral symptoms of ASD (Navarro et al. 2016; Slattery et al. 2016). Probiotics are believed to improve GIT symptoms, yet a review by Critchfield et al. (2011) also suggests the role of high levels of *Clostridium* species in triggering GIT problems in children with ASD. Among their many benefits, the authors noted that probiotics can stabilize the mucosal barrier, reduce bacterial overgrowth, synthesize antioxidants, increase the intestinal permeability, and stimulate immunity in ASD.

By therapeutically administering propionic acid and clindamycin in an animal model, probiotic treatment has shown to reduce autistic-like excitation/inhibition imbalance and ameliorate glutamate excitotoxicity by increasing depleted GABA and Mg2+ and decreasing glutamate (El-Ansary et al. 2018). In children, probiotic supplementation resulting in increased colony counts of Bifidobacteria and Lactobacilli levels has been associated with a significant reduction in body weight and clinical improvements in the severity of ASD and gastrointestinal symptoms (Shaaban et al. 2018).

6.7.2 Attention-Deficit Hyperactivity Disorder (ADHD)

Attention-deficit hyperactivity disorder (ADHD) is a neurodevelopmental disease characterized into different representations: inattentiveness, impulsivity, and/or hyperactivity or all three characteristics combined. ADHD prevalence is estimated to be 5.3–7.2% in children under 18, which makes it the most common neurobehavioral disorder in children. Variations in diagnosis usually cause differences in the estimations of the prevalence rather than geographic variations. ADHD can still present in adulthood as 30–60% of diagnosed children continue to have symptoms in their adulthood with inattentive type being the most predominant. The

Therapeutically, the use of vitamin D supplementation in children with ADHD through the administration of 50,000 IU/week vitamin was reported to decrease the severity of ADHD symptoms. The administration of 2000 IU/day besides methylphenidate was also reported to improve evening ADHD symptoms in children.

6.8 Conclusion

This chapter aimed to provide a summary of the nutrigenomics of neurodevelopmental disorders. A review of research suggests nutrition, including prenatal exposure to nutrients, either as risk or protective factors with significant interactions with genes. Preliminary findings suggest the involvement of specific factors such as vitamin D, gene polymorphism, iron deficiency, prenatal exposure, and dietary variables as potential mediators. More research is required to better understand the mechanisms of action by which nutrition and genes interact in cases of neurodevelopmental disorders and the clinical implications of such findings in terms of guiding effective treatment.

References

Aarts E, Ederveen THA, Naaijen J, Zwiers MP, Boekhorst J, Timmerman HM, Smeekens SP, Netea MG, Buitelaar JK, Franke B et al (2017) Gut microbiome in ADHD and its relation to neural reward anticipation. PLoS One 12:e0183509

Adams JB, Johansen LJ, Powell LD, Quig D, Rubin RA (2011) Gastrointestinal flora and gastrointestinal status in children with autism—comparisons to typical children and correlation with autism severity. BMC gastroenterology 11:22

Ali A, Cui X, Eyles D (2018) Developmental vitamin D deficiency and autism: putative pathogenic mechanisms. J Steroid Biochem Mol Biol 175:108–118. https://doi.org/10.1016/j.jsbmb.2016.12.018

Ali A, Vasileva S, Langguth M, Alexander S, Cui X, Whitehouse A, McGrath JJ, Eyles D (2019) Developmental vitamin D deficiency produces behavioral phenotypes of relevance to autism in an animal model. Nutrients 11(5):1187. https://doi.org/10.3390/nu11051187

Antshel KM, Zhang-James Y, Faraone SV (2013) The comorbidity of ADHD and autism spectrum disorder. Expert Rev Neurother 13(10):1117–1128. https://doi.org/10.1586/14737175.2013.840417

Balta B, Gumus H, Bayramov R, Korkmaz Bayramov K, Erdogan M, Oztop DB, Dogan ME, Taheri S, Dundar M (2018) Increased vitamin D receptor gene expression and rs11568820 and rs4516035 promoter polymorphisms in autistic disorder. Mol Biol Rep 45(4):541–546. https://doi.org/10.1007/s11033-018-4191-y

Bölte S, Girdler S, Marschik PB (2019) The contribution of environmental exposure to the etiology of autism spectrum disorder. Cell. Mol. Life Sci 76:1275–1297. https://doi.org/10.1007/s00018-018-2988-4

Bourgeois JP (1997) Synaptogenesis, heterochrony and epigenesis in the mammalian neocortex. Acta Paediatr Suppl 422:27–33

Buie T, Campbell DB, Fuchs GJ et al (2010) Evaluation, diagnosis, and treatment of gastrointestinal disorders in individuals with ASDs: a consensus report. Pediatrics 125:S1–S18

Bull-Larsen S, Mohajeri MH (2019) The potential influence of the bacterial microbiome on the development and progression of ADHD. Nutrients 11(11):2805. https://doi.org/10.3390/nu11112805

Cannell JJ (2008) Autism and vitamin D. Med Hypotheses 70:750–759

Cannell JJ (2017) Vitamin D and autism, what's new? Rev Endocr Metab Disord 18:183–193. https://doi.org/10.1007/s11154-017-9409-0

Chmielewska A, Dziechciarz P, Gieruszczak-Białek D, Horvath A, Pieścik-Lech M, Ruszczyński M, Skórka A, Szajewska H (2019) Effects of prenatal and/or postnatal supplementation with iron, PUFA or folic acid on neurodevelopment: update. Br J Nutr 122(s1):S10–S15. https://doi.org/10.1017/S0007114514004243

Cenit MC, Nuevo IC, Codoñer-Franch P, Dinan TG, Sanz Y (2017) Gut microbiota and attention deficit hyperactivity disorder: new perspectives for a challenging condition. Eur Child Adolesc Psychiatry 26(9):1081–1092. https://doi.org/10.1007/s00787-017-0969-z

Codagnone MG, Stanton C, O'Mahony SM, Dinan TG, Cryan JF (2019) Microbiota and neurodevelopmental trajectories: role of maternal and early-life nutrition. Ann Nutr Metab 74 (Suppl 2):16–27. https://doi.org/10.1159/000499144

Coury DL, Ashwood P, Fasano A et al (2012) Gastrointestinal conditions in children with autism spectrum disorder: developing a research agenda. Pediatrics 130(2):S160–S168

Cerdó T, Diéguez E, Campoy C (2019) Early nutrition and gut microbiome: interrelationship between bacterial metabolism, immune system, brain structure, and neurodevelopment. Am J Physiol Endocrinol Metab 317(4):E617–E630. https://doi.org/10.1152/ajpendo.00188.2019

Critchfield JW, van Hemert S, Ash M, Mulder L, Ashwood P (2011) The potential role of probiotics in the management of childhood autism spectrum disorders. Gastroenterol Res Pract

D'Ambrosio A, Pagani E, Riccitelli GC, Colombo B, Rodegher M, Falini A, Comi G, Filippi M, Rocca MA (2017) Cerebellar contribution to motor and cognitive performance in multiple sclerosis: an MRI sub-regional volumetric analysis. Mult Scler 23:1194–1203

Darling AL, Rayman MP, Steer CD, Golding J, Lanham-New SA, Bath SC (2017) Association between maternal vitamin D status in pregnancy and neurodevelopmental outcomes in childhood: results from the Avon Longitudinal Study of Parents and Children (ALSPAC). Br J Nutr 117:1682–1692

Dean JC, Hailey H, Moore SJ, Lloyd DJ, Turnpenny PD, Little J (2002) Long term health and neurodevelopment in children exposed to antiepileptic drugs before birth. J Med Genet 39:251–259

Dhamayanti M, Noviandhari A, Supriadi S, Judistiani RT, Setiabudiawan B (2019) Association of maternal vitamin D deficiency and infants' neurodevelopmental status: a cohort study on vitamin D and its impact during pregnancy and childhood in Indonesia. J Paediatr Child Health. https://doi.org/10.1111/jpc.14481

Dosman C, Brian J, Drmic I, Senthilselvan A, Harford M, Smith R et al (2007) Children with autism: effect of iron supplementation on sleep and ferritin. Pediatr Neurol 36(3):152–158. https://doi.org/10.1016/j.pediatrneurol.2006.11.004

Dufault R, Schnoll R, Lukiw WJ, LeBlanc B, Cornett C, Patrick L, Wallinga D, Gilbert SG, Crider R (2009) Mercury exposure, nutritional deficiencies and metabolic disruptions may affect learning in children. Behav Brain Funct 5:44–10. https://doi.org/10.1186/1744-9081-5-44

El-Ansary A, Bacha AB, Bjørklund G et al (2018) Probiotic treatment reduces the autistic-like excitation/inhibition imbalance in juvenile hamsters induced by orally administered propionic acid and clindamycin. Metab Brain Dis 33:1155–1164. https://doi.org/10.1007/s11011-018-0212-8

Eyles DW, Feron F, Cui X, Kesby JP, Harms LH, Ko P et al (2009) Developmental vitamin D deficiency causes abnormal brain development. Psychoneuroendocrinology 34(Suppl 1):S247–S257

Eyles DW, Burne THJ, McGrath JJ (2013) Vitamin D, effects on brain development, adult brain function and the links between low levels of vitamin D and neuropsychiatric disease. Front Neuroendocrinol 34:47–64

FAO/WHO (2001) Health and nutrition properties of probiotics in food including powder milk with live lactic acid bacteria. Report of a Joint FAO/WHO expert consultation on evaluation of health and nutritional properties of probiotics in food including powder milk with live lactic acid bacteria. FASEB J 19:1067–1077. https://doi.org/10.1096/fj.04-3284com. 64

Feng J, Shan L, Du L, Wang B, Li H, Wang et al (2017) Clinical improvement following vitamin D3 supplementation in Autism Spectrum Disorder. Nutr Neurosci 20(5):284–290. https://doi.org/10.1080/1028415X.2015.1123847

Fox SE, Levitt P, Nelson CA III (2010) How the timing and quality of early experiences influence the development of brain architecture. Child Dev 81(1):28–40

Gernand AD, Schulze KJ, Stewart CP, West KP Jr, Christian P (2016) Micronutrient deficiencies in pregnancy worldwide: health effects and prevention. Nat Rev Endocrinol 12(5):274–289. https://doi.org/10.1038/nrendo.2016.37

Gilmour A (1938) The mental condition in rickets. School Hygiene 1912(9):6–16

Gogou M, Kolios G (2018) Are therapeutic diets an emerging additional choice in autism spectrum disorder management? World J Pediatr 14(3):215–223. https://doi.org/10.1007/s12519-018-0164-4

Gombart AF, Borregaard N, Koeffler HP (2005) Human cathelicidin antimicrobial peptide (CAMP) gene is a direct target of the vitamin D receptor and is strongly up-regulated in myeloid cells by 1,25-dihydroxyvitamin D3. FASEB J 19(9):1067–1077. https://doi.org/10.1096/fj.04-3284com

Grecksch G, Ruthrich H, Hollt V, Becker A (2009) Transient prenatal vitamin D deficiency is associated with changes of synaptic plasticity in the dentate gyrus in adult rats. Psychoneuroendocrinology 34(Suppl 1):S258–S264

Hallerhan MM (1938) The effect of rickets on the mental development of young children. Arch Psychol 229:1–67

Huang YN, Ho YJ, Lai CC, Chiu CT, Wang JY (2015) 1,25-dihydroxyvitamin D3 attenuates endotoxin-induced production of inflammatory mediators by inhibiting MAPK activation in primary cortical neuron-glia cultures. J Neuroinflammation 12:147

Huttenlocher PR, Dabholkar AS (1997) Regional differences in synaptogenesis in human cerebral cortex. J Comp Neurol 387(2):167–178

Javadfar Z, Abdollahzad H, Moludi J, Rezaeian S, Amirian H, Foroughi AA, Nachvak SM, Goharmehr N, Mostafai R (2020) Effects of vitamin D supplementation on core symptoms, serum serotonin, and interleukin-6 in children with autism spectrum disorders: a randomized clinical trial. Nutrition:79–80. https://doi.org/10.1016/j.nut.2020.110986

Kerley CP, Power C, Gallagher L et al (2017) Lack of effect of vitamin D3 supplementation in autism: a 20-week, placebo-controlled RCT. Arch Dis Child 102:1030–1036

Kraeuter AK, Guest PC, Sarnyai Z (2019) The therapeutic potential of ketogenic diet throughout life: focus on metabolic, neurodevelopmental and neurodegenerative disorders. Adv Exp Med Biol 1178:77–101. https://doi.org/10.1007/978-3-030-25650-0_5

Kočovská E, Fernell E, Billstedt E et al (2012) Vitamin D and autism: clinical review. Res Dev Disabil 33(5):1541–1550

Kotsi E, Kotsi E, Perrea DN (2019) Vitamin D levels in children and adolescents with attention-deficit hyperactivity disorder (ADHD): a meta-analysis. Attent Defic Hyperact Disord 11(3): 221–232. https://doi.org/10.1007/s12402-018-0276-7

Kostović I, Judas M, Rados M, Hrabac P (2002) Laminar organization of the human fetal cerebrum revealed by histochemical markers and magnetic resonance imaging. Cereb Cortex 12(5): 536–544

Levenson CW, Figueiroa SM (2008) Gestational vitamin D deficiency: long-term effects on the brain. Nutr Rev 66:726–729

Levitt P (2003) Structural and functional maturation of the developing primate brain. Pediatr 143(4 suppl):S35–S45

Li YJ, Li YM, Xiang DX (2018) Supplement intervention associated with nutritional deficiencies in autism spectrum disorders: a systematic review. Eur J Nutr 57(7):2571–2582. https://doi.org/10.1007/s00394-017-1528-6

Lozoff B (2011) Early iron deficiency has brain and behavior effects consistent with dopaminergic dysfunction. J Nutr 141(Suppl):S740–S746

Lou H (1994) Dopamine precursors and brain function in phenylalanine hydroxylase deficiency. Acta Paediatr 83:86–88

Ly V, Bottelier M, Hoekstra PJ, Arias Vasquez A, Buitelaar JK, Rommelse NN (2017) Elimination diets' efficacy and mechanisms in attention deficit hyperactivity disorder and autism spectrum disorder. Eur Child Adolesc Psychiatry 26(9):1067–1079. https://doi.org/10.1007/s00787-017-0959-1

Matson JL, Shoemaker M (2009) Intellectual disability and its relationship to autism spectrum disorders. Res Dev Disabil 30(6):1107–1114. https://doi.org/10.1016/j.ridd.2009.06.003

Milman N, Taylor CL, Merkel J, Brannon PM (2017) Iron status in pregnant women and women of reproductive age in Europe. Am J Clin Nutr 106(Suppl 6):1655S–1662S. https://doi.org/10.3945/ajcn.117.156000. Epub 2017 Oct 25. PMID: 29070543; PMCID: PMC5701710

Mobasheri L, Moossavi SZ, Esmaeili A, Mohammadoo-khorasani M, Sarab GA (2020) Association between vitamin D receptor gene FokI and TaqI variants with autism spectrum disorder predisposition in Iranian population. Gene 723. https://doi.org/10.1016/j.gene.2019.144133

Moore SJ, Turnpenny P, Quinn A et al (2000) A clinical study of 57 children with fetal anticonvulsant syndromes. J Med Genet 37:489–497

Navarro F, Liu Y, Rhoads JM (2016) Can probiotics benefit children with autism spectrum disorders? World J Gastroenterol 22:10093–10098

Ogbu D, Xia E, Sun J (2020) Gut instincts: vitamin D/vitamin D receptor and microbiome in neurodevelopment disorders. Open Biol 10(7):200063. https://doi.org/10.1098/rsob.200063

Ornoy A (2009) Valproic acid in pregnancy: how much are we endangering the embryo and fetus? Reprod Toxicol 28:1–10

Pärtty A, Kalliomäki M, Wacklin P et al (2015) A possible link between early probiotic intervention and the risk of neuropsychiatric disorders later in childhood: a randomized trial. Pediatr Res 77:823–828. https://doi.org/10.1038/pr.2015.51

Patrick RP, Ames BN (2014) Vitamin D hormone regulates serotonin synthesis. Part 1: relevance for autism. FASEB J 28:2398–2413. https://doi.org/10.1096/fj.13-246546

Petra AI, Panagiotidou S, Hatziagelaki E, Stewart JM, Conti P, Theoharides TC (2015) Gut–microbiota–brain axis and its effect on neuropsychiatric disorders with suspected immune dysregulation. Clin Ther 37(5):984–995

Piwowarczyk A, Horvath A, Łukasik J, Pisula E, Szajewska H (2018) Gluten- and casein-free diet and autism spectrum disorders in children: a systematic review. Eur J Nutr 57(2):433–440. https://doi.org/10.1007/s00394-017-1483-2

Pivina L, Semenova Y, Doşa MD, Dauletyarova M, Bjørklund G (2019) Iron deficiency, cognitive functions, and neurobehavioral disorders in children. J Mol Neurosci 68(1):1–10. https://doi.org/10.1007/s12031-019-01276-1

Pojednic RM, Ceglia L, Olsson K, Gustafsson T, Lichtenstein AH, Dawson-Hughes B, Fielding RA (2015) Effects of 1,25-dihydroxyvitamin D3 and vitamin D3 on the expression of the vitamin D receptor in human skeletal muscle cells. Calcif Tissue Int 96:256–263. https://doi.org/10.1007/s00223-014-9932-x

Prosser DE, Jones G (2004) Enzymes involved in the activation and inactivation of vitamin D. Trends Biochem Sci 29:664–673. https://doi.org/10.1016/j.tibs.2004.10.005

Ristori MV, Quagliariello A, Reddel S, Ianiro G, Vicari S, Gasbarrini A, Putignani L (2019) Autism, gastrointestinal symptoms and modulation of gut microbiota by nutritional interventions. Nutrients 11(11):2812. https://doi.org/10.3390/nu11112812

Saad K, Abdel-Rahman AA, Elserogy YM, Al-Atram AA, El-Houfey A, Othman HA, Bjørklund G, Jia F, Urbina MA, Abo-Elela MGM, Ahmad F-A, Abd El-Baseer KA, Ahmed AE, Abdel-Salam AM (2018) Retracted: randomized controlled trial of vitamin D supplementation in children with autism spectrum disorder. J Child Psychol Psychiatr 59:20–29. https://doi.org/10.1111/jcpp.12652

Schmidt RJ, Hansen RL, Hartiala J, Allayee H, Sconberg JL, Schmidt LC, Volk HE, Tassone F (2015) Selected vitamin D metabolic gene variants and risk for autism spectrum disorder in the CHARGE study. Early Hum Dev 91(8):483–489

Schmidt RJ, Hansen RL, Hartiala J et al (2011) Prenatal vitamins, one-carbon metabolism gene variants, and risk for autism. Epidemiology 22:476–485

Shaaban SY, El Gendy YG, Mehanna NS et al (2018) The role of probiotics in children with autism spectrum disorder: a prospective, open-label study. Nutr Neurosci 21(9):676–681. https://doi.org/10.1080/1028415x.2017.1347746

Slattery J, MacFabe DF, Frye RE (2016) The significance of the enteric microbiome on the development of childhood disease: a review of prebiotic and probiotic therapies in disorders of childhood. Clin Med Insights Pediatr 10:91–107

Song L, Luo X, Jiang Q, Chen Z, Zhou L, Wang D, Chen A (2020) Vitamin D supplementation is beneficial for children with autism spectrum disorder: a meta-analysis. Clin Psychopharmacol Neurosci 18(2):203–213. https://doi.org/10.9758/cpn.2020.18.2.203

Stiles J, Jernigan TL (2010) The basics of brain development. Neuropsychol Rev 20(4):327–348

Tan M, Yang T, Zhu J, Li Q, Lai X, Li Y, Tang T, Chen J, Li T (2020) Maternal folic acid and micronutrient supplementation is associated with vitamin levels and symptoms in children with autism spectrum disorders. Reproduct toxicol(Elmsford, NY) 91:109–115. https://doi.org/10.1016/j.reprotox.2019.11.009

Van De Sande MM, van Buul VJ, Brouns FJ (2014) Autism and nutrition: the role of the gut-brain axis. Nutr Res Rev 27(2):199–214. https://doi.org/10.1017/S0954422414000110

van Sadelhoff J, Perez Pardo P, Wu J, Garssen J, van Bergenhenegouwen J, Hogenkamp A, Hartog A, Kraneveld AD (2019) The gut-immune-brain axis in autism spectrum disorders; a focus on amino acids. Front Endocrinol 10:247. https://doi.org/10.3389/fendo.2019.00247

Vela G, Stark P, Socha M, Sauer AK, Hagmeyer S, Grabrucker AM (2015) Zinc in gut-brain interaction in autism and neurological disorders. Neural Plast:1–15. https://doi.org/10.1155/2015/972791

Villamor E, Rifas-Shiman SL, Gillman MW et al (2012) Maternal intake of methyl-donor nutrients and child cognition at 3 years of age. Paediatr Perinat Epidemiol 26:328–335

Vohr BR, Davis EP, Wanke CA, Krebs NF (2017) Neurodevelopment: the impact of nutrition and inflammation during preconception and pregnancy in low-resource settings. Pediatrics 139 (Suppl 1):S38–S49. https://doi.org/10.1542/peds.2016-2828F

Vuillermot S, Luan W, Meyer U et al (2017) Vitamin D treatment during pregnancy prevents autism-related phenotypes in a mouse model of maternal immune activation. Mol Autism 8:9. https://doi.org/10.1186/s13229-017-0125-0

Wiegersma AM, Dalman C, Lee BK, Karlsson H, Gardner RM (2019) Association of prenatal maternal anemia with neurodevelopmental disorders. JAMA Psychiat 76(12):1294–1304. https://doi.org/10.1001/jamapsychiatry.2019.2309

Yu H, Zhang Z, Liu J et al (2020) Association study between genetic variants in vitamin D metabolism related genes and childhood autism spectrum disorder. Metab Brain Dis 35:971–978. https://doi.org/10.1007/s11011-020-00570-x

Zhong W, Gu B, Gu Y, Groome LJ, Sun J, Wang Y (2014) Activation of vitamin D receptor promotes VEGF and CuZn-SOD expression in endothelial cells. J Steroid Biochem Mol Biol 140:56–62. https://doi.org/10.1016/j.jsbmb.2013.11.017

Zou R, El Marroun H, McGrath JJ, Muetzel RL, Hillegers M, White T, Tiemeier H (2020) A prospective population-based study of gestational vitamin D status and brain morphology in preadolescents. NeuroImage 209:116514. https://doi.org/10.1016/j.neuroimage.2020.116514

Chapter 7
Nutrition and Alzheimer's Disease

Sara Ayman Moustafa, Nouran Yonis, and Mohamed Salama

7.1 Introduction

Dementia is a condition in which memory, perception, behavior, and the capacity to do normal tasks deteriorate. While dementia affects the elderly predominantly, it is not a common aspect of aging. Around 50 million people have dementia globally, with almost 60% residing in low- and middle-income countries, and almost 10 million new cases are reported per year (Dementia 2020). There are about 10 million new cases each year. The average percentage at a given time of the general population aged 60 and over with dementia is between 5% and 8%. It is estimated that the overall number of people with dementia will exceed 82 million in 2030 and 152 million in 2050 (Dementia 2020). Dementia in older people worldwide is one of the main causes of disease and dependence. Dementia not only has a physical, mental, social, and economic effect on individuals with dementia but also on their relatives, friends, and community as a whole (Dementia 2020).

While age is the most established dementia risk factor, it is not an expected consequence of aging. In addition, dementia does not only affect elderly adults, accounting for up to 9% of cases of young-onset dementia (defined as the onset of symptoms before the age of 65 years). Studies suggest that by having enough exercise, not smoking, limiting unhealthy use of alcohol, managing their weight, consuming a balanced diet, and keeping healthy levels of blood pressure,

S. A. Moustafa (✉) · N. Yonis
Institute of Global Health and Human Ecology, School of Sciences and Engineering, The American University of Cairo, New Cairo, Egypt
e-mail: saraayman@aucegypt.edu

M. Salama
Institute of Global Health and Human Ecology, School of Sciences and Engineering, The American University of Cairo, New Cairo, Egypt

Faculty of Medicine, Manousra University, Mansoura, Egypt

M. Salama (ed.), *Nutrigenomics and the Brain*, Nutritional Neurosciences,
https://doi.org/10.1007/978-981-16-9205-5_7

cholesterol, and blood sugar, people can decrease their risk of dementia. Depression, low educational achievement, social exclusion, and cognitive inactivity are additional risk factors (Dementia 2020). No therapy is currently available to reverse or change the gradual course of dementia. In separate phases of clinical trials, several new therapies are being studied. The primary dementia treatment priorities are encouraging early and optimal management; optimizing physical fitness, cognition, exercise, and well-being; and detecting and treating physical, behavioral, and psychological symptoms.

In terms of direct medical and social care costs and the cost of informal services, dementia has substantial social and economic consequences. The overall global social burden of dementia was estimated at US$ 818 billion in 2015, representing 1.1% of the world's gross domestic product. As a proportion of GDP, the overall expense ranged from 0.2% in low- and middle-income countries to 1.4% in high-income countries. For the family of disabled individuals and for their caregivers, dementia may be daunting. Physical, emotional, and financial stresses may place tremendous burden on families and caregivers, and health, social, financial, and legal frameworks need help (Dementia 2020).

The most prevalent cause of dementia is Alzheimer's disease which leads to 60–70% of cases (Dementia 2020). Alzheimer's disease is a progressive neurological disease resulting in loss of neurons and their connections. Such changes affect the person's ability to remember, think, and live independently. Scientists still don't know the exact cause for Alzheimer's disease. Genetic, environmental, and lifestyle factors and changes in the brain are some of the suggested causes that vary from an individual to another (CDC 2020).

In 2014, there were as many as 5 million Americans with Alzheimer's disease. Over the age of 60, the symptoms of the disease may first appear, and the risk increases with age. Alzheimer's disease may occur in younger adults, although it is less prevalent. Beyond the age of 65, the number of individuals living with the condition doubles every 5 years. By 2060, this figure is estimated to almost triple to 14 million persons (CDC 2020).

There are two types of Alzheimer's disease—early-onset and late-onset. Both types have a genetic component. Late onset is the most common type. Signs first appear in a person's mid-60s which may involve a gene called APOE ε4. On the other hand, the early-onset Alzheimer is very rare to occur. Signs first appear between a person's 30s and mid-60s and usually caused by gene changes passed down from parent to offspring (CDC 2020).

Nutritious diet, physical activity, social engagement, sleep, and mentally stimulating pursuits have all been associated with helping people stay healthy as they age which may decrease the risk of Alzheimer's disease. Ongoing researches are trying to understand the causes of Alzheimer's disease to develop better treatment and prevention approaches. One suggested approach is nutrigenomics and the role they play in the brain and development of Alzheimer's disease.

7.2 Etiology of Alzheimer's Disease

According to The Centers for Disease Control and Prevention (CDC 2020), scientists are still trying to find out the exact cause for Alzheimer's disease. There is definitely not a single factor, but multiple variables influence each person differently. Physical, mental, and social activities may reduce the risk of Alzheimer's disease, yet some factors may increase the risk of AD such as the following:

- Age
- Family history
- Changes in the brain years before the first symptoms appear
- Education, diet, and environment
- Heart disease and stroke, such as high blood pressure and high cholesterol

7.3 Role of Nutrition in Alzheimer's Disease

AD is a complex neurodegenerative disease where early diagnosis and treatment are still not of great challenge. It was crucial to find alternative pathways and techniques that can help with the delay of disease onset as well as early detection and diagnosis. The role of lifestyle and nutrition were widely discussed. A connection was found between a healthy lifestyle and delayed disease onset; nutrition was found to have a dynamic role in the prevention and delaying the disease. Various nutritional approaches were discussed in the literature for nutrition and AD. AB and tau proteins are the major proteins involved with AD; their accumulation leads to plaque formation and progress of the disease. Hyperphosphorylation of tau protein was reported to cause microtubule disassembly and neuronal loss in the memory and learning centers in the brain. Another disease etiology is the dysregulation of mitochondria, where its vital role in metabolizing nutrients is compromised in AD patients. An alteration in the biogenesis and dynamics of mitochondria are landmarks of AD where the oxidative stress is increased by the excess generation of free radicals with age.

Nutritional approaches were taken to counteract this damaging effect of AD on vital body systems and organelles. Most micronutrients (vitamins, minerals, essential amino acids, and essential fatty acids, including omega-3 polyunsaturated fatty acids (PUFAs)) have been studied in relation to cerebral functioning. Thiamine (vitamin B1), which is essential for the maintenance of the brain function, when in decreased supply, results in different disorders of the nervous system (both peripheral and central nervous system defects). Calorie restriction, antioxidant-rich diet, and certain dietary approaches like adopting a ketogenic diet, Mediterranean diet, and other diets were found to limit the progress of metabolic and neurodegenerative diseases. These dietary interventions were reported to be able to modify the epigenetic mechanism of AD by regulating DNA methylation, histone modifications, acetylation, and miRNA expression which can influence the gene expression responsible for epigenetic alterations.

7.4 Gut-Brain Axis in AD

The role of microbiome and gut microbiota has gained a lot of attention lately due to the evident link between the gut and brain (gut-brain axis) and its effect on neurodegenerative disease through modulating neurochemical and neurometabolic pathway between CNS and ENS including communication between endocrine complex immune systems; it was reported that dysbiosis and alteration in the gut microbiota can play a vital role in early diagnosis and early prevention of neurodegenerative diseases through interference with synthesis and secretion of neurotrophic factors, for example, brain-derived neurotrophic factor and gamma-aminobutyric acid and NMDA receptors (Junges et al. 2018). Gut dysbiosis and alteration in the gut microbiota induce increased permeability of gut barrier (leaky gut) and immune activation which can lead to systemic inflammation that in turn can impair the blood-brain barrier and promote neuroinflammation, neural injury, and neurodegeneration. The elderly are prone to persistent inflammatory state of the gut which can have an impact on microglia maturation, which is attributed to short-chain fatty acids that are products of bacterial metabolism. Microbiota influences peripheral immune cell activation and cytokine profile which cause CNS inflammation. In the enteric nervous system, the neurons are organized in a microcircuits manner that allows the GIT to function independently from CNS although the systems are interconnected and influence one another. GI dysfunction in the APP mice showed an accumulation of AB in the enteric neurons which lead to a decrease in the number of enteric neurons as well as its dysmotility; these effects will lead to increased vulnerability to inflammation. Curli amyloid bacteria are best studied in AD. These bacterial amyloids share multiple similarities to CNS amyloids. Exposure to bacterial amyloid can cause priming of the immune system which in turn improves the immune response to endogenous production of neuronal amyloid in the brain. Chen et al. reported increased deposition of neuronal alpha synuclein in the gut and brain accompanied by enhanced microgliosis and astrogliosis. Animal models were used to validate the gut microbiota role in AD. Minter et al. reported that antibiotic-induced perturbations in the gut microbiota induce neuroinflammation and amyloidosis in a rat model of AD (Kowalski and Mulak 2019).

7.5 Diets Role in AD

Prevention and treatment of AD are still not fully established; thus, other alternatives were investigated that can possess protective and preventive effects in AD patients. Increased attention was drawn lately towards the gut-brain axis and how dietary options and various diets can play a role in AD. Various diets were implemented, and their effects were very promising as protective agents and in slowing the progression of the disease.

7.5.1 Ketogenic Diet

A correlation was established between Alzheimer's disease and apolipoprotein E and ApoE4 genetic variants specifically as it is known as the Alzheimer's disease gene. People with the ApoE4 gene are at higher risk of AD and other physiological consequences. Using nutrition as an alternative intervention for AD was discussed in the literature. Carbohydrate-restricted high-fat ketogenic diet is one of the approaches used in AD treatment. Ketogenic diet was thought to have an impact not only on AD but also on metabolic syndrome in patients. When implemented, a decrease in HOMA-IR, triglycerides, VLDL, and HdA1c was reported with an improvement in MoCA score. Following a ketogenic diet was promising for AD patients with ApoE4 variant, as it can yield an overall improvement in metabolic health and cognition through restoring metabolic flexibility to the brain and peripheral tissues. Dietary ketosis is reported to work by regulating the kinase nutrient signaling pathway of Mtok/AMPK which occurs through ketone body production where it works as a fuel substrate to the starving brain in AD patients and restore metabolic pathology in dysregulated neurons (Morrill and Gibas 2019).

7.5.2 Intermittent Fasting

Another approach for AD prevention through nutrition and dietary routine is intermittent fasting. Shin et al. (2018) studied the effect of intermittent fasting on rats and its effect on metabolism and cognitive functions. It was reported that intermittent fasting decreases serum glucose levels by increasing insulin secretion, as well as improving dyslipidemia and liver damage index compared to non-AD rats. AD not only impairs cognitive function, but it affects the overall metabolic activity from distributed energy, lipid, and bone metabolism. Intermittent fasting was found to protect against alteration in these metabolic parameters as well as improve cognitive functions in AD patients.

7.5.3 Mediterranean Diet

Several researches were conducted and found a positive correlation between Mediterranean diet and AD. The Mediterranean diet is based on food that contains vitamins, polyphenols, and unsaturated fatty acids which could be beneficial in

Alzheimer's disease prevention (Miranda et al. 2017; Yusufov et al. 2017). One recent study discussed the modified Mediterranean-ketogenic diet and the importance of fatty acids in the prevention of Alzheimer's disease (Nagpal et al. 2019; Román et al. 2019). Further research should be directed toward the importance of Mediterranean diet in the prevention of Alzheimer's disease.

7.6 Role of Vitamins in AD

Vitamins could play a significant role in the prevention of Alzheimer's disease. Many research studies were implemented to prove this. The most studied vitamins that showed promising effects are vitamins B, D, and E as discussed below.

7.6.1 Vitamin B

Since the role of nutrition have been investigated and considered in the prevention of AD, vitamins have been of great importance especially vitamins of the B group and their role as antioxidants in gene control for endothelium protection as a coenzyme where they take part in the biochemical reactions in the brain where they were reported for their cooperation with other elements which is vital for brain health (An et al. 2019) (Szczechowiak et al. 2019). They take part in the synthesis of polyunsaturated phosphatidylcholine by cooperating with choline through S-adenosylmethionine (SAM) methyl donation. Their anti-inflammatory properties were discussed as a protective role against AD through modulation of calcium and glutamate currents.

Vitamin B1 (thiamine) is a labile quaternary ammonia product found in unphosphorylated thiamine. It provides a refining system to protect the hippocampal neurons cultured with glutamate excess. Thiamine deficiency induces an excess of glutamate release diminished by glutamatergic blocking action with NMDA antagonist. These findings are yet to be applied in clinical practices (Boston et al. 2020).

Vitamin B2 (riboflavin) is mainly obtained from dietary intake in which its active form is riboflavin or the coenzyme flavin mononucleotide (FMN) and flavin adenine dinucleotide (FAD). It possess a neuroprotective function as an antioxidant where it is involved in glutathione reduction. It was reported through animal studies that vitamin B2 plays a vital role in reducing the oxidative damage after reperfusion through its direct action on free radicals as shown in animal studies (D'Cunha et al. 2019). It was also reported that it has a role in the reduction of glutamate excitotoxicity through glutamate neuronal release inhibition. Vitamin B2 deficiency through inadequate dietary intake was related with cognitive decline. However, these studies are yet to be proven and implemented in clinical practices in AD (McCleery et al. 2018).

Vitamin B3 (niacin) consists of three molecules, namely, niacin, nicotinamide, and nicotinamide riboside. When coupled with tryptophan, they are used for the synthesis of NAD+ and NADPH. They are crucial for mitochondrial respiration and ATP formation. Another role for the vitamin B3 family is promoting calcium signaling and acting as neurotransmitter through the purine receptor. Vitamin B3 can be obtained from various dietary sources. However, its deficiency can have serious implications as the lack of NAD causes pellagra (McCleery et al. 2018). Its role in the brain is studied as a cell protector, where it protects neurons from axonal degradation derived from excitotoxicity. In AD its main effect is on cerebral microvascular endothelial cells and neurons. It protects these cells from ischemic insults and oxidative damage and influences inflammatory processes related to microglial functions. Vitamin B3 supplements were tested on animals and humans with AD which showed promising results; in animal testing, it was reported to prevent AB accumulation and hippocampal astrocyte loss, thus improving cognitive impairment. However, these results were not replicated when done on humans with AD yet. More clinical studies are needed to validate the use of vitamin B3 in AD patients (An et al. 2019; Szczechowiak et al. 2019).

7.6.2 Vitamin D

One of the implications of AD is hippocampal neurogenesis impairment found at the early stages of AD which cause early cognitive decline. Vitamin D was investigated for its link to the pathophysiology of AD and to hippocampal neurogenesis. Vit D effect was examined in animal model of AD to examine the effect of vitamin D. It was reported that vitamin D when delivered reduces cognitive decline during the symptomatic phase. It was reported that early hypovitaminosis D would cause an increase in the amyloid plaque in AD mice, while late hypovitaminosis D caused an impairment in neurogenesis in mice. Tromso's study reported that vitamin D can be used as a predictive of cognitive outcome in the older population, as age can play as a threshold for the implications of vitamin D in cognitive functions. Another longitudinal study reported that hypovitaminosis D can be predictive of AD at younger ages. Further studies are necessary to better understand the impact of vitamin D and its use in AD treatment.

7.6.2.1 Impact of Oxidative Stress and Mitochondrial Dysfunction

AD is accompanied with mitochondrial function as it affects metabolism, energy production, free radical production, and calcium buffering and increases oxidative stress and affects activation of protease and phospholipases. Agnihotri and Aruoma (2020) reported that maintaining antioxidant systems as superoxide dismutase and glutathione peroxidase is essential for the normal function of mitochondria, limiting the amount of oxidative damage. Vitamin E, vitamin C, and CoQ10 (a coenzyme that transfers electrons from complexes I and II to complex III, optimizing the proton gradient of the oxidative phosphorylation mechanisms) are used for this purpose since antioxidants in addition to these vitamins in diet have shown positive impact on AD-induced rats but still has limited efficacy in AD patients as these antioxidants are unstable chemicals. Targeting mitochondrial interaction and modifying abnormal mitochondrial function can be a promising approach for AD patients.

7.6.3 Vitamin E

Vitamin E has gained a lot of attention recently with its effect on neurodegenerative diseases. The two isoforms of vitamin E tocopherol and tocotrienol possess various properties, namely, anti-inflammatory, antioxidant, lowering cholesterol, and cellular signaling. Those properties can play a vital role in AD. Vitamin E concentrations were found to decrease, closer to diminished in AD patients. Vitamin E intake may limit disease progression. This is due to the synergistic effect found between vitamin E isoform and other micronutrients. Clinical evidence is still limited to validate the usefulness of vitamin E intake in AD patients.

7.7 Other Nutrients

7.7.1 Homocysteine

Homocysteine plays a vital role in the body as it is strongly correlated with B12, folate, and B6 vitamin deficiency (Zaric et al. 2019), as hyperhomocysteinemia is a direct indicator of alteration in these vitamins in the body. Homocysteine was also found to have a role in Alzheimer's disease and its prodromal stage where it is considered a risk factor for AD and mild cognitive impairment. Various studies were conducted over the years to investigate the role of homocysteine and its role in AD, where it was established that high serum levels of Hcy are correlated with early onset of AD. The mechanism by which HCY levels are correlated to AD was explained through tissue remodeling through metalloproteinases MMPs where Hcy causes a disruption in the blood-brain barrier and causes microvascular leak through MMP9 action (Zhang et al. 2017). Another explanation was related to folate deficiency that caused Hcy-dependent inhibition of methylation of phosphatase, causing phosphatase activity inhibition that in turn causes an increase in phospho-tau immunoreactivity. Vitamins B supplementations were found to improve Hcy serum levels, which reflects on slowing down cognitive decline progress. The main limitations of studying Hcy is the lack of enough human studies that validate the use of vitamin B supplements to reverse hyperhomocysteinemia. Ultimately, understanding the underlying mechanism of hyperhomocysteinemia in AD beyond vitamin B cofactors might help better optimize treatment strategies, which may result in direct clinical benefit to patients (Bhargava et al. 2018; Zhang et al. 2017).

7.7.2 *Selenium*

Selenium (Se), an important trace factor, is vital to the brain, but, based on dose and speciation, it can be possibly neurotoxic; Se has been discussed in relation to Alzheimer's disease for decades (Solovyev et al. 2018). Selenoprotein P (SELENOP) is a heparin-binding secreted glycoprotein that in mammals acts as the primary Se transport protein. Studies in vivo found that this protein could have additional roles, such as a contribution to redox control. Recent literature found a potential role of SELENOP in AD pathology, based on models and human trials (Solovyev et al. 2018). SELENOP plays a role in the activity of the brain, such as its engagement in glial stimulation and its contribution to both "healthy" and pathological signaling pathways in neuronal and glial tissue, including its role in homeostasis/excitotoxicity of neuronal calcium. Thus, human studies should pay more attention to the role of Se (Solovyev et al. 2018).

A review study was conducted by Varikasuvu et al. (2019) comparing the tissue levels of AD and non-AD brains of selenium. The results revealed that selenium levels in AD were decreased in temporal, hippocampal, and cortex regions. Further studies need to confirm this association.

7.7.3 *Nutraceuticals and AD*

Compared to the single-target activity of most medications used for AD care, nutraceutical compounds offer the benefit of a multitarget strategy, labelling various molecular sites in the human brain (Camila et al. 2020). These targets include acetylcholinesterase suppression, β-amyloid senile plaques, products for oxidation, inflammatory processes, particular brain receptors, etc. and pharmacological events as varied as anti-inflammatory, memory enhancing, nootropic effects, excitotoxicity of glutamate, antidepressants, and antioxidants (Camila et al. 2020). Examples for nutraceuticals include curcumin, morin, luteolin, resveratrol, and many others. One successful approach is the use of BrainUp-10 nutraceutical formulations consisting of Andean Shilajit and vitamin B complex, with memory-enhancing activity and neuropsychiatric distress management in patients with AD (Camila et al. 2020). This applied nutraceutical opens a new direction for future research and clinical trials that are expected to deliver such findings based on medical evidence.

7.7.4 *Polyphenols*

Polyphenols are antioxidants that were found to have a role in AD through modulating the gut-brain axis where these polyphenolic compounds are transformed into physiologically active and neuroprotector compounds that are then referred to as human lignans; this process is done through gut bacterial metabolism. These gut bacterial metabolites were reported to have a neuroprotective effect in different neurodegenerative diseases like AD, where enterolactone and enterodiol are secondary gut bacterial metabolites that are formed act as acetylcholinesterase inhibitors and can then be used as therapeutic and protective agents in AD. Another protective role of polyphenols is exerted by advanced glycation end product (AGE) inhibitors where they are antiglycating agents (Freyssin et al. 2018). Another interesting role of polyphenols in AD is the ability of polyphenolic compounds as hesperetin, hesperidin, and neohesperidin to cross the blood-brain barrier. Animal studies on AD mice showed that gut bacterial metabolism of flavan-3-ols will lead to the formation of aryl-y-valerolactone and arylvaleric acid derivatives. These metabolites and their secondary metabolites were found to selectively detoxify AB oligomers, thus preventing memory loss in mouse AD models. Further metabolism of valerolactones produces other polyphenolic products as hydroxyarylcinnamic acid, hydroxy arylpropionic acid, hydroxybenzoic acid, and others. These secondary metabolites were found to have higher bioavailability and permeability to BBB than dietary flavonoids, thus attenuating neuroinflammation (Reddy et al. 2020).

Polyphenolic lignin compounds are found in flaxseeds and other dietary plans with fruits, wheat, and vegetables. The gut bacteria is responsible for demethylation and deglycosylation of these plant-based lignans to produce human lignans. Increasing polyphenolic lignans in the diet of AD patients will result in the inhibition of acetylcholinesterase from a natural source, thus protecting and preventing disease progression (Freyssin et al. 2018; Reddy et al. 2020).

7.8 Treatment

Many treatment approaches are used to treat Alzheimer's disease. Most of the patients are treated using medications, yet the ideal approach should include changing dietary habits. As discussed in the book chapter, different diets and nutrients can have a positive impact in Alzheimer's disease patients. Ongoing research on new approaches showed promising results. For example, fatty acids are one of the most important nutrient elements in the human brain, involved in the supply of energy, cell structure, and metabolism pathways. For brain lipid homeostasis and neurodegenerative disease prevention, safe eating patterns and adequate fatty acid consumption are helpful (Zhang et al. 2020). Another example is based on the zinc element. Zinc plays an important role in the brain and the central nervous system (Bitanihirwe and Cunningham 2009; Frederickson 1989). Researchers recently found the use of zinc transport proteins which can play a role in curing Alzheimer's disease (Xu et al. 2019), and also probiotic and prebiotic might improve mental function in Alzheimer's disease patients (Ansari et al. 2020).

In conclusion, further research is needed. In addition, we should pay more attention to the possible molecular pathways of fatty acids, zinc transporter, and probiotic in AD anatomy, finding more information and goals in AD therapy and drug development in regard to the possible drug interactions that may occur between nutrients and medications administered to AD patients.

7.9 Future Perspectives

In this book chapter, we aimed to highlight the importance of nutrients in the management plan of AD patients. We suggest that further research should be directed toward the new approaches that showed promising results in curing AD. We believe that nutrition is a very important aspect in the management plan in patients diagnosed with AD. More clinical and longitudinal studies should be conducted to validate the promising role of nutrition as a protective agent and in delaying disease progression. Validation of specific elimination diets or other dietary regimens is crucial so that it can be implemented in AD treatment protocols.

References

Agnihotri A, Aruoma OI (2020) Alzheimer's disease and Parkinson's disease: a nutritional toxicology perspective of the impact of oxidative stress, mitochondrial dysfunction, nutrigenomics and environmental chemicals. J Am Coll Nutr 39(1):16–27. https://doi.org/10.1080/07315724.2019.1683379

An Y, Feng L, Zhang X, Wang Y, Wang Y, Tao L, Qin Z, Xiao R (2019) Dietary intakes and biomarker patterns of folate, vitamin B6, and vitamin B12 can be associated with cognitive impairment by hypermethylation of redox-related genes NUDT15 and TXNRD1. Clin Epigenetics 11(1):139. https://doi.org/10.1186/s13148-019-0741-y

Ansari F, Pourjafar H, Tabrizi A, Homayouni A (2020) The effects of probiotics and prebiotics on mental disorders: a review on depression, anxiety, Alzheimer, and autism Spectrum disorders. Curr Pharm Biotechnol 21(7):555–565. https://doi.org/10.2174/1389201021666200107113812

Bhargava S, Bhandari A, Choudhury S (2018) Role of homocysteine in cognitive impairment and Alzheimer's disease. Indian J Clin Biochem 33(1):16–20. https://doi.org/10.1007/s12291-017-0646-5

Bitanihirwe BK, Cunningham MG (2009) Zinc: the brain's dark horse. Synapse (New York, NY) 63(11):1029–1049. https://doi.org/10.1002/syn.20683

Boston PF, McKirdy SJ, Al-Turki MA, Barker ME, Russell JM (2020) Vitamin B12 and folate levels in progression of Alzheimer's disease - a short report. Int J Psychiatry Clin Pract 24(1):68–70. https://doi.org/10.1080/13651501.2019.1681457

CDC (2020) What is Alzheimer's disease? Retrieved December 15, 2020, from https://www.cdc.gov/aging/aginginfo/alzheimers.htm

Camila C et al (2020) The emerging role of nutraceuticals and phytochemicals in the prevention and treatment of Alzheimer's disease. J Alzheimers Dis 77:33–51

D'Cunha NM, Georgousopoulou EN, Boyd L, Veysey M, Sturm J, O'Brien B, Lucock M, McKune AJ, Mellor DD, Roach PD, Naumovski N (2019) Relationship between B-vitamin biomarkers and dietary intake with apolipoprotein E ϵ4 in Alzheimer's disease. J Nutr Gerontol Geriatr 38(2):173–195. https://doi.org/10.1080/21551197.2019.1590287

Dementia (2020) https://www.who.int/news-room/fact-sheets/detail/dementia. Accessed 15 Dec 2020

Frederickson CJ (1989) Neurobiology of zinc and zinc-containing neurons. Int Rev Neurobiol 31: 145–238. https://doi.org/10.1016/s0074-7742(08)60279-2

Freyssin A, Page G, Fauconneau B, Rioux Bilan A (2018) Natural polyphenols effects on protein aggregates in Alzheimer's and Parkinson's prion-like diseases. Neural Regen Res 13(6): 955–961. https://doi.org/10.4103/1673-5374.233432

Junges VM, Closs VE, Nogueira GM, Gottlieb MGV (2018) Crosstalk between gut microbiota and central nervous system: a focus on Alzheimer's disease. Curr Alzheimer Res 15(13): 1179–1190. https://doi.org/10.2174/1567205015666180904155908

Kowalski K, Mulak A (2019) Brain-gut-microbiota Axis in Alzheimer's disease. J Neurogastroenterol Motil 25(1):48–60. https://doi.org/10.5056/jnm18087

Miranda A, Gómez-Gaete C, Mennickent S (2017) Dieta mediterránea y sus efectos benéficos en la prevención de la enfermedad de Alzheimer [Role of Mediterranean diet on the prevention of Alzheimer disease]. Rev Med Chil 145(4):501–507. Spanish. https://doi.org/10.4067/S0034-98872017000400010

McCleery J, Abraham RP, Denton DA, Rutjes AW, Chong LY, Al-Assaf AS, Griffith DJ, Rafeeq S, Yaman H, Malik MA, Di Nisio M, Martínez G, Vernooij RW, Tabet N (2018) Vitamin and mineral supplementation for preventing dementia or delaying cognitive decline in people with mild cognitive impairment. Cochrane Database Syst Rev 11(11):CD011905. https://doi.org/10.1002/14651858.CD011905.pub2

Morrill SJ, Gibas KJ (2019) Ketogenic diet rescues cognition in ApoE4+ patient with mild Alzheimer's disease: a case study. Diabetes Metab Syndr 13(2):1187–1191. https://doi.org/10.1016/j.dsx.2019.01.035

Nagpal R, Neth BJ, Wang S, Craft S, Yadav H (2019) Modified Mediterranean-ketogenic diet modulates gut microbiome and short-chain fatty acids in association with Alzheimer's disease markers in subjects with mild cognitive impairment. EBioMedicine 47:529–542. https://doi.org/10.1016/j.ebiom.2019.08.032

Reddy VP, Aryal P, Robinson S, Rafiu R, Obrenovich M, Perry G (2020) Polyphenols in Alzheimer's disease and in the gut-brain Axis. Microorganisms 8(2):199. https://doi.org/10.3390/microorganisms8020199

Román GC, Jackson RE, Gadhia R, Román AN, Reis J (2019) Mediterranean diet: the role of long-chain ω-3 fatty acids in fish; polyphenols in fruits, vegetables, cereals, coffee, tea, cacao and wine; probiotics and vitamins in prevention of stroke, age-related cognitive decline, and Alzheimer disease. Rev Neurol 175(10):724–741. https://doi.org/10.1016/j.neurol.2019.08.005

Shin BK, Kang S, Kim DS, Park S (2018) Intermittent fasting protects against the deterioration of cognitive function, energy metabolism and dyslipidemia in Alzheimer's disease-induced estrogen deficient rats. Exp Biol Med (Maywood) 243(4):334–343. https://doi.org/10.1177/1535370217751610

Solovyev N, Drobyshev E, Bjørklund G, Dubrovskii Y, Lysiuk R, Rayman MP (2018) Selenium, selenoprotein P, and Alzheimer's disease: is there a link? Free Radic Biol Med 127:124–133. https://doi.org/10.1016/j.freeradbiomed.2018.02.030

Szczechowiak K, Diniz BS, Leszek J (2019) Diet and Alzheimer's dementia—nutritional approach to modulate inflammation. Pharmacol Biochem Behav 184:172743. https://doi.org/10.1016/j.pbb.2019.172743

Varikasuvu SR, Prasad VS, Kothapalli J, Manne M (2019) BRAIN selenium in Alzheimer's disease (BRAIN SEAD study): a systematic review and meta-analysis. Biol Trace Elem Res 189(2): 361–369. https://doi.org/10.1007/s12011-018-1492-x

Xu Y, Xiao G, Liu L, Lang M (2019) Zinc transporters in Alzheimer's disease. Mol Brain 12(1): 106. https://doi.org/10.1186/s13041-019-0528-2

Yusufov M, Weyandt LL, Piryatinsky I (2017) Alzheimer's disease and diet: a systematic review. Int J Neurosci 127(2):161–175. https://doi.org/10.3109/00207454.2016.1155572

Zaric BL, Obradovic M, Bajic V, Haidara MA, Jovanovic M, Isenovic ER (2019) Homocysteine and Hyperhomocysteinaemia. Curr Med Chem 26(16):2948–2961. https://doi.org/10.2174/0929867325666180313105949

Zhang DM, Ye JX, Mu JS, Cui XP (2017) Efficacy of vitamin B supplementation on cognition in elderly patients with cognitive-related diseases. J Geriatr Psychiatry Neurol 30(1):50–59. https://doi.org/10.1177/0891988716673466

Zhang T, Han X, Zhang X, Chen Z, Mi Y, Gou X (2020) Dietary fatty acid factors in Alzheimer's disease: a review. J Alzheimer's Dis: JAD 78(3):887–904. https://doi.org/10.3233/JAD-200558

Chapter 8
Nutrigenomics in Parkinson's Disease

Jihan Azar

8.1 Introduction

Parkinson's disease (PD) is the second most common neurodegenerative disease in the world (Ball et al. 2019). PD is characterized by the progressive loss of dopaminergic neurons in a brain region called the basal ganglia, more specifically the substantia nigra par compacta (Nasuti et al. 2017). Lewy bodies are the main proteins which deposit in the brains of PD patients (Tysnes and Storstein 2017). These proteins are composed of alpha-synuclein protein. There are many factors playing a role in the pathogenesis of PD. Ten to fifteen percent of PD cases are familial, while the rest are sporadic, a product of gene environmental interaction (Ball et al. 2019; Nasuti et al. 2017).

PD cardinal symptoms include tremor, rigidity, and bradykinesia (Nasuti et al. 2017; Tysnes and Storstein 2017). Non-motor symptoms of PD include anosmia, constipation, or depression. These may precede the motor symptoms by years (Nasuti et al. 2017).

This chapter will talk about the effect of nutrition on PD, in terms of genetic variation among patients. First, we will talk about the gut-brain axis, which is the first link between nutrition and the brain, and then we will go into detail, talking about each gene known to be implicated in PD and how some specific food affect this gene and its functional pathway.

J. Azar (✉)
Institute of Global Health and Human Ecology (IGHHE), The American University in Cairo (AUC), New Cairo, Egypt
e-mail: jihan.azar@aucegypt.edu

M. Salama (ed.), *Nutrigenomics and the Brain*, Nutritional Neurosciences,
https://doi.org/10.1007/978-981-16-9205-5_8

8.1.1 Gut-Brain Axis

The gut-brain axis is an important part of PD, as it strengthens the link between nutrition and the development of disease and its presentation.

Lewy bodies, which are aggregates of alpha-synuclein proteins, are found in the dorsal motor nucleus and olfactory bulb and anterior olfactory nucleus of PD patients (Braak et al. 2003). They were also found in the gastrointestinal tract of PD patients in the prodromal phase of the disease (nearly 10–20 years before the diagnosis) (Stokholm et al. 2016). They are also spread along the spinal cord, autonomic ganglia, adrenal medulla, submandibular gland, heart, and enteric nervous system (Goedert et al. 2017). This might explain the reason why PD patients mostly die from pneumonia and cardio- and cerebrovascular complications.[34] The presence of alpha-synuclein protein aggregations may be enough to cause functional deficits in the enteric, peripheral, and central nervous system in the absence of cell death (Peelaerts and Baekelandt 2016). α-synuclein aggregates were proven to hasten neurodegeneration (Goedert et al. 2017). The pathway through which alpha-synuclein proteins travel whether from the CNS to the ENS or vice versa is still vaguely known.

The vagus nerve is the tenth cranial nerve and is the most important parasympathetic nerve in our body. The dorsal nucleus in the medulla serves as its nucleus. The preganglionic parasympathetic vagal neurons are very long, and the parasympathetic ganglia are proximal to their target organs (Feher 2016). Most vagal fibers connecting the gut and the brain are afferent.[24] The intestine is connected to the brain via the enteric nervous system. The ENS is comprised of two major ganglia, responsible for motor and secretory functions: the myenteric (Auerbach's) plexus in the muscularis propria layer and the submucous (Meissner's) plexus (Costa et al. 2000). Both receive messages from the vagus nerve and responds accordingly. The myenteric plexus extends its nerves from the upper esophagus to the internal anal sphincter, and the submucosal plexus is found in the small and large intestine (Furness et al. 2014).

Toxins/cytokines from gut microorganisms can travel through the epithelial lining of the gastrointestinal tract and reach postganglionic parasympathetic fibers which are then connected to the preganglionic parasympathetic motor neurons of the vagal nerve after transsynaptic transmission from the axon of Meissner's plexus. The end route for the vagus is the dorsal medial nucleus which is located in the medulla (Hawkes et al. 2009). Short-chain fatty acids (SCFA) such as butyric acid can be easily transported across the epithelial barrier and activate vagal chemoreceptors. This is one of the communication pathways that could play a role in the PD pathogenesis (Forsythe et al. 2014).

The mechanism by which alpha-synuclein aggregates are transported could be exocytosis from neurons and extracellular alpha-synuclein endocytosis via neurons, astrocytes, and microglia (Lee et al. 2011). Chronic inflammation was also shown to contribute to the pathogenesis of PD. Alpha-synuclein overexpression in genetically engineered mice allowed for lipopolysaccharide-induced inflammation in dopaminergic neurons and more alpha-synuclein aggregation in the nigrostriatal neurons (Gao et al. 2008). Another hypothesis for alpha-synuclein migration could be the microtubule associated transport (Furness et al. 2014).

Poor diet in the elderly is known to cause low microbiota diversity, inflammation, and disability (Quigley 2017). Due to evolution, our diet and environment led to a change in our microbiota, not in favor of our immune regulation, predisposing us to disease (Forsythe et al. 2014). This is especially true in PD. In this chapter we will focus on the effect of nutrients on PD through genetics.

8.2 State of the Art

8.2.1 PD Pathogenesis: Different Pathways Affected

Oxidative stress has been demonstrated in playing a role in the pathogenesis of neurodegenerative diseases such as PD (Agnihotri and Aruoma 2020). Some of the changes oxidative stress induces include mitochondrial DNA mutations, depression of the mitochondrial respiratory chain functions, alteration of membrane permeability, disturbance of Ca2+ homeostasis, and damage to mitochondrial antioxidant system (Agnihotri and Aruoma 2020). One of the theories is that reactive oxygen species overproduction in mitochondria can oxidize thiol groups in proteins and subsequently promote calcium uptake. Calcium overload can cause cell death in these cells (Agnihotri and Aruoma 2020). The reason behind the overproduction of reactive oxygen species is postulated to be mitochondrial DNA mutation (Agnihotri and Aruoma 2020).

8.2.2 Genetics and PD

The most common gene mutation in PD are found in genes such as alpha-synuclein (SNCA), glucocerebrosidase (GBA), leucine-rich repeat kinase 2 (LRRK2), vacuolar protein sorting-associated protein 35 (VPS35), parkin RBR E3 ubiquitin protein ligase (PARK2), phosphatase and tensing homolog-induced kinase 1 (PINK1), and Parkinson protein 7 (PARK7) (Ball et al. 2019).

8.2.2.1 PARK Genes

These genes are responsible for about 5–10% of PD cases (Park and Ellis 2020). Mutations in these genes are implicated in oxidative stress; they either increase the production of ROS or reduce their clearance. These lead to the production of abnormal proteins (Park and Ellis 2020).

8.2.2.2 Alpha-Synuclein (SNCA) or PARK1 Gene

In normal conditions, alpha-synuclein protein helps in synaptogenesis and synaptic plasticity (Park and Ellis 2020). In PD, this protein aggregates and presents the main neuropathological finding by forming Lewy bodies. These aggregates contribute to the increased ROS production and lipid peroxidation. They also affect the mitochondrial membrane potential and open mitochondrial death channels (Park and Ellis 2020).

8.2.2.3 PARK2 Gene

Regulation of the mitochondrial quality control is done by parkin. Parkin is a ubiquitin E3 ligase which is encoded by PARK2 gene. PARK2 gene mutations are the most common causes of early-onset PD. Parkin mutation itself accounts for 77% of all cases under 30 years of age (Park and Ellis 2020).

8.2.2.4 PARK6 Gene

PARK6 gene is responsible for encoding PTEN-induced kinase 1 (PINK1) which is an enzyme responsible for autophagy. Parkin and PINK1 work together to remove dysfunctional mitochondria (mitophagy). Any mutation or deletion of these two will impair mitophagy, therefore resulting in failure to remove dysfunctional mitochondria and increasing ROS production and glutathione depletion (Park and Ellis 2020).

8.2.2.5 PARK7 Gene

DJ-1 protein encoded by PARK7 gene has an important role in mitochondrial function. DJ-1 protein binds to ATP synthase and the antiapoptotic protein Bcl-xL promoting the survival of dopaminergic neurons. Depletion of DJ-1 exposes mitochondria to oxidative stress, whereas its overexpression serves as a protection for neurons due to more intracellular antioxidant production (Park and Ellis 2020). The posttranslational modifications of this protein during oxidative stress include oxidation at cysteine-106 and then translocation to mitochondria, thereby preventing its damage (Park and Ellis 2020).

Tripterygium wilfordii is a Chinese herb also known under the name thunder god vine. It has antioxidant properties because it contains the compound triterpene. The administration of this herb to flies who were deficient in DJ-1 gene showed a promising effect on the dopaminergic neurons and dopamine concentration in the brain (Evangelakou et al. 2019).

8.2.2.6 PARK8 Gene/LRRK2 Gene

Dardarin or LRRK2 is a kinase found in the mitochondrial membrane and interacts with other PARK gene products such as parkin, PINK1, and DJ-1. Mutations in the PARK8 gene increases the kinase activity. This is common in late-onset autosomal-dominant PD. Mutations in the G2019S domain of the LRRK2 gene increases ROS-induced dopaminergic neuronal death. LRRK2 inhibitors decrease ROS in the same way, restoring mitochondrial function and decreases proapoptotic proteins including caspase 3, Bax, and apoptotic-inducing factor (Park and Ellis 2020). Thymoquinone is a kinase inhibitor which affects fly models with G2019S mutations in the LRRK gene. It has been shown to lower the prooxidant effects of the G2019S mutation, in addition, reducing the loss of dopaminergic neurons and locomotor defects (Evangelakou et al. 2019). On a transcriptomic level, a decrease in gene expression which supports synaptic activity and mitochondrial function has been seen in the striatum of PD patients. Downregulation of immune function genes and upregulation of genes implicated in inflammation signaling, oxidative stress, and DNA repair have also been noted in PD (Joseph Lawrence Webb 2020).

8.2.3 Epigenetics and PD

Epigenetic mechanisms such as DNA methylation, histone modification, and non-coding microRNAs play a role in the response of the body to nutrition. Nutrition can cause epigenetic changes by directly influencing gene expression, activating nuclear receptors by ligands, and modifying of membrane receptor signaling cascades. Epigenetic mechanisms are robust and temporal; they correlate with the development of the organism, are tissue/organ specific, and responsive to the internal and external environment (Virmani et al. 2013).

8.2.4 Diet and PD

In a survey-based study, PD severity was lower in patients consuming specific food like fresh vegetables, fresh fruit, nuts and seeds, fish, olive oil, wine, coconut oil, fresh herbs, and the use of spices (Mischley et al. 2017). Fruits and vegetables contain flavonoids and carotenoids which reduce the risk of neurodegenerative diseases (Agnihotri and Aruoma 2020). The presence of vitamins such as vitamin E, vitamin C, and vitamin A have polyphenols and anthocyanins which also contribute to the neuroprotective effects (Agnihotri and Aruoma 2020).

The Mediterranean diet has been associated with lower PD incidence and later age at diagnosis (Mischley et al. 2017).

8.2.4.1 Dairy

Dairy food, on the other hand, has been associated with higher rates of PD progression and higher PD incidence (Mischley et al. 2017). It has to be noted that not all kinds of dairy food act the same on the pathogenesis of PD. There have been many studies assessing this risk, but the results have been inconsistent (Mischley et al. 2017). Mechanisms behind this pathogenesis may include uric acid reduction due to dairy intake, which is known to hasten PD progression and is association with higher PD incidence. This is by deleting the peroxynitrite in the CNS (Mischley et al. 2017). Another mechanism can be by increasing insulin resistance due to dairy consumption or lactose intolerance which increases with age. The latter mechanism might be due to increased intestinal inflammation and permeability which occurs when there's a lack of lactase enzyme in the presence of dairy intake. People from African, Asian, Hispanic, and Native American decent more commonly present with lactose intolerance. The last two mechanisms are neurotoxic contaminants present in dairy products, such as pesticides, or bovine microbiota being introduced and affecting the intestinal microflora (Mischley et al. 2017).

8.2.4.2 Alcohol

Red wine is reported to be protective of PD when consumed moderately (Agnihotri and Aruoma 2020) or in high amounts (Fall et al. 1999) because of the presence of polyphenols, among other substances, which modulates nuclear erythroid 2-related factors which improve cognitive functions and also act as antioxidants (Agnihotri and Aruoma 2020). Coffee, beer, wines, and liquor all reduce the risk of neurodegenerative diseases according to Fall et al. (Fall et al. 1999).

8.2.4.3 Fruits, Vegetables, and Vitamins

Canned fruits and vegetables were associated with PD progression; the main reason behind was postulated to be either bisphenol A (BPA) which is used to coat the inner of the cans or the neurotoxicant aluminum in the cans (Mischley et al. 2017).

Vitamin B5 or pantothenic acid is an important micronutrient for neurotransmitter synthesis as well as oxidative metabolism of lipids and carbohydrates. It's deficiency is linked to symptoms such as neuropathy. Thereby vitamin B5 can affect neuronal function by impairing neurons' cellular metabolism and synthesis of neurotransmitters (Virmani et al. 2013).

Green tea is regarded as an antioxidant due to its content of catechins (of the family polyphenols). It exerts its neuroprotective effect by protecting the mitochondrial function/cellular energy production (Virmani et al. 2013).

8.2.4.4 Coenzyme Q10

Coenzyme Q10 supplement was associated with reduced rates of PD. Clinical trials have nevertheless controversial results, as some show improvement in motor function of patients, and other associations were explained when accounting for income measures (Mischley et al. 2017).

8.2.4.5 Fish Oil and Seafood

Fish oil was associated with lower PD rates. It contains omega-3 fatty acids such as eicosapentaenoic acid (EPA) and docosahexaenoic acid (DHA). DHA has neuroprotective and antioxidant effects and additionally is a precursor for neuroprotection, which exerts antiapoptotic activity. It also enhances synaptogenesis, neurogenesis, and neurite outgrowth and promotes the synthesis of brain-derived neurotrophic factor (BDNF) (Mischley et al. 2017). DHA and EPA are also found in seafood and are known to maintain the structural integrity of neuronal membranes (Agnihotri and Aruoma 2020). They also promote excitability of neuron cell membranes, reduce inflammation, and suppress the production of free radicals (Włodarek 2019). This might play an important role in neurodegenerative diseases.

8.2.4.6 Iron

Iron is needed for dopamine synthesis in the substantia nigra. But excess iron may expose this region to higher vulnerability; therefore, most studies are consistent with the fact that iron is associated with PD progression. Red meat is one of the foods containing high amounts of iron, which correlated with PD progression when compared with other meals (Mischley et al. 2017).

8.2.4.7 Middle Eastern Diet

A Greek study showed that prodromal PD in elderly was less common among those adhering to Mediterranean diet (Agnihotri and Aruoma 2020).

8.2.4.8 Fried Foods

Lipid peroxidation and subsequent increase in reactive oxygen species are common with fried foods. Reactive oxygen species are also commonly seen among PD patients. In addition, lipid peroxidation yields the by-product of aldehydes, for example, acrolein, which binds to thiol groups on proteins yielding their aggregating. Specifically in PD, acrolein accumulates in dopaminergic neurons, modifies alpha-synuclein, and inhibits proteasome activity (Mischley et al. 2017).

8.2.4.9 Ketogenic Diet

The ketogenic diet is a kind of diet where the individuals take in a high amount of fat and low amounts of carbohydrates to induce a state of ketosis in the body. The brain relies on glucose as the main source of energy. If glucose is not available to the brain for 2–3 days, either through glycogenesis or food, an alternative energy source will be identified. This source is the ketone bodies, acetoacetate, 3-hydroxybutyrate, and acetone. Ketogenesis happens in the mitochondria of the liver (Włodarek 2019).

Oxidative stress (production of reactive oxygen species) is implicated in the pathogenesis of PD by dysregulating dopamine metabolism (Włodarek 2019).

The ketogenic diet has an antioxidant effect by increasing glutathione and glutathione peroxidase activity. This was observed in rats and might be potentially protective of the neurodegeneration in the hippocampus (Włodarek 2019).

Dopaminergic neuron degeneration in the substantia nigra which occurs in PD was replicated in rats by applying MPTP to them. The ketone body, beta-hydroxyacetate, was shown to be neuroprotective in this case. Octanoic acid had the same effect by increasing the metabolic activity in the striatal mitochondria. Ketogenic diet was also able to modify motor function in PD rats for the better (Włodarek 2019).

Ketones also have an effect on the synaptic functionality. In addition, they act as antioxidative agents, improve ATP synthesis, affect the ATP-sensitive potassium channel (KATP), and reduce the dopaminergic cell death due to neurotoxicant exposure via glutathione activity (Włodarek 2019).

PD patients who adhered to the ketogenic diet for 28 days showed some improvement in the UPDRS Scale (Włodarek 2019). It should be noted that PD patients on levodopa will experience an increase in its bioavailability if they had a low-protein diet (as seen in the ketogenic diet), as levodopa and proteins compete for intestinal absorption (Mischley et al. 2017; Włodarek 2019). Ketogenic diet was also associated with improvement in non-motor symptoms (Włodarek 2019).

8.2.4.10 Caloric Intake

Lower calorie intake was associated with lower risk of neurodegenerative diseases (Agnihotri and Aruoma 2020). Caloric restriction is neuroprotective by reducing the reactive oxygen species production, increasing neuroprotective factors (BDNF, NT-3, GDNF) and molecular chaperones.

It can also exert anti-inflammatory effects by reducing the NFaB levels, TNFa, COX-2, and interleukins such as IL1B, IL2, IL4, and IL6 (Włodarek 2019).

8.2.4.11 Sugar

Diet soda in specific and soda in general was also found to have a faster progression rate of PD. Soda contributes to high caloric intake and obesity (Mischley et al. 2017).

An increase in aspartic acid and phenylamine after the consumption of some sugars might interfere with the transport of serotonin and dopamine to the brain which will increase the brain's excitability and lead to degeneration in astrocytes and neurons (Mischley et al. 2017).

8.2.4.12 Meat

Alpha-synuclein is the protein found in the Lewy bodies which are the hallmark of PD. They are also found in the enteric nervous system where they are thought to activate immune cells. Both milk and meat may possess some proteins that will have cross-reactivity with other antigens in our enteric nervous system. Both milk and meat have been associated with incidence and progression of PD. Both being high in proteins can affect levodopa absorption, so the effect should be studied more deeply to determine whether it's the medication absorption or other constituents that affect the pathogenesis of the disease (Mischley et al. 2017).

8.2.4.13 Eggs

Egg consumption was associated with a lower risk of PD in a Swedish population who eat eggs every day. This relationship was consistent in other populations such as Chinese where subjects who had more than five eggs per day had a lower relative risk; nevertheless it didn't reach statistical significance. PD symptom severity was also reported to be lower with the consumption of eggs (Joseph Lawrence Webb 2020).

8.2.4.14 Miscellaneous

Gene expression networks controlling circadian rhythm are upregulated in PD patients (Joseph Lawrence Webb 2020). Melatonin is an antioxidant which exerts neuroprotective effects. This might be also linked to the circadian rhythm (Virmani et al. 2013).

Impaired mitochondrial function is one of the important defects found in neurodegenerative diseases such as PD (Joseph Lawrence Webb 2020). So any nutrient which preserves mitochondrial function will act as a neuroprotective agent. These include alpha-lipoic acid, coenzyme Q10, carnitines, and nicotinamide (Virmani et al. 2013). Carnitine have a specific mechanism of action in which they act as neuroprotective agents. It acts via gene modulation, inhibiting NrF2 transcription factors, reducing oxidative stress (Virmani et al. 2013).

References

Agnihotri A, Aruoma OI (2020) Alzheimer's disease and Parkinson's disease: a nutritional toxicology perspective of the impact of oxidative stress, mitochondrial dysfunction, nutrigenomics and environmental chemicals. J Am Coll Nutr. https://doi.org/10.1080/07315724.2019.1683379

Ball N, Teo W-P, Chandra S, Chapman J (2019) Parkinson's disease and the environment. Front Neurol. https://doi.org/10.3389/fneur.2019.00218

Braak H, Del Tredici K, Rüb U et al (2003) Staging of brain pathology related to sporadic Parkinson's disease. Neurobiol Aging. https://doi.org/10.1016/S0197-4580(02)00065-9

Costa M, Brookes SJH, Hennig GW (2000) Anatomy and physiology of the enteric nervous system. Gut. https://doi.org/10.1136/gut.47.suppl_4.iv15

Evangelakou Z, Manola M, Gumeni S, Trougakos IP (2019) Nutrigenomics as a tool to study the impact of diet on aging and age-related diseases: the drosophila approach. Genes Nutr 14:12

Fall PA, Fredrikson M, Axelson O, Granérus AK (1999) Nutritional and occupational factors influencing the risk of Parkinson's disease: a case-control study in southeastern Sweden. Mov Disord. https://doi.org/10.1002/1531-8257(199901)14:1<28::AID-MDS1007>3.0.CO;2-O

Feher J (2016) Quantitative human physiology: an introduction. Quantitative Human Physiology

Forsythe P, Bienenstock J, Kunze WA (2014) Vagal pathways for microbiome-brain-gut axis communication. Adv Exp Med Biol. https://doi.org/10.1007/978-1-4939-0897-4_5

Furness JB, Callaghan BP, Rivera LR, Cho HJ (2014) The enteric nervous system and gastrointestinal innervation: integrated local and central control. Adv Exp Med Biol. https://doi.org/10.1007/978-1-4939-0897-4_3

Gao HM, Kotzbauer PT, Uryu K et al (2008) Neuroinflammation and oxidation/nitration of α-synuclein linked to dopaminergic neurodegeneration. J Neurosci. https://doi.org/10.1523/JNEUROSCI.0143-07.2008

Goedert M, Masuda-Suzukake M, Falcon B (2017) Like prions: the propagation of aggregated tau and α-synuclein in neurodegeneration. Brain 140(2):266–278

Hawkes CH, Del Tredici K, Braak H (2009) Parkinson's disease: the dual hit theory revisited. Ann N Y Acad Sci 1170:615–622

Joseph Lawrence Webb (2020) Nutrigenomics and aging: a transcriptomic perspective. ProQuest Dissertations and Theses

Lee H-J, Suk J-E, Lee K-W et al (2011) Transmission of Synucleinopathies in the enteric nervous system of A53T alpha-Synuclein transgenic mice. Exp Neurobiol. https://doi.org/10.5607/en.2011.20.4.181

Mischley LK, Lau RC, Bennett RD (2017) Role of diet and nutritional supplements in Parkinson's disease progression. Oxid Med Cell Longev. https://doi.org/10.1155/2017/6405278

Nasuti C, Brunori G, Eusepi P et al (2017) Early life exposure to permethrin: a progressive animal model of Parkinson's disease. J Pharmacol Toxicol Methods. https://doi.org/10.1016/j.vascn.2016.10.003

Park HA, Ellis AC (2020) Dietary antioxidants and Parkinson's disease. Antioxidants 9(7):570

Peelaerts W, Baekelandt V (2016) α-Synuclein strains and the variable pathologies of synucleinopathies. J Neurochem 139(Suppl 1):256–274

Quigley EMM (2017) Microbiota-brain-gut Axis and neurodegenerative diseases. Curr Neurol Neurosci Rep 17(12):–94

Stokholm MG, Danielsen EH, Hamilton-Dutoit SJ, Borghammer P (2016) Pathological α-synuclein in gastrointestinal tissues from prodromal Parkinson disease patients. Ann Neurol. https://doi.org/10.1002/ana.24648

Tysnes OB, Storstein A (2017) Epidemiology of Parkinson's disease. J Neural Transm 124(8): 901–905

Virmani A, Pinto L, Binienda Z, Ali S (2013) Food, nutrigenomics, and neurodegeneration—neuroprotection by what you eat! Mol Neurobiol 48(2):353–362

Włodarek D (2019) Role of ketogenic diets in neurodegenerative diseases (Alzheimer's disease and Parkinson's disease). Nutrients 11(1):169

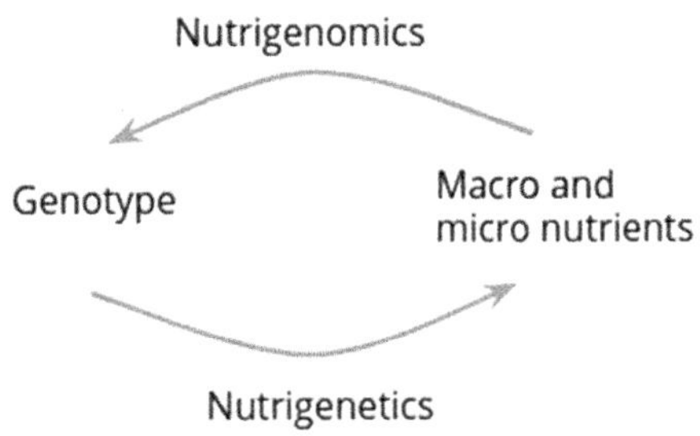

Fig. 9.1 Nutrigenetics and nutrigenomics are based on the bidirectional interaction between cellular/genetic processes, nutrients, and nutritional genetics

two parts: nutrigenomics and nutrigenetic (Ordovas and Corella 2004; Kaput and Rodriguez 2004).

The importance of nutritional genetics stems from the potential to develop a holistic understanding of how different dietary nutrients can either increase or decrease the risk of disease by promoting damaging DNA or aid in disease treatment by maintaining the stability of DNA. Understanding the ability of how certain individuals' intake and metabolize certain nutrients compared to others has helped in further understanding the etiology of diseases.

The study of nutritional genetics and its two parts (nutrigenomics and nutrigenetics) could potentially help in developing dietary patterns, functional food, and supplementation to improve genome health that is personalized for individuals (Paoloni-Giacobino et al. 2003; Bull and Fenech 2008).

The conceptual basis and the tenets of nutritional genetics were defined by both Jim Kaput and Raymond L. Rodriguez in their paper "Nutritional Genomics: The Next Frontier in the Postgenomic Era" as the following(Kaput and Rodriguez 2004):

1. Common dietary chemicals act on the human genome, either directly or indirectly, to alter gene expression and/or structure.
2. Under certain circumstances and in some individuals, diet can be a serious risk factor for a number of diseases.
3. Some diet-regulated genes, including common variants, are likely to play a role in the onset, incidence, progression, and/or severity of chronic diseases.
4. The degree to which diet influences the balance between healthy and disease states may depend on an individual's genetic background.
5. Dietary intervention based on knowledge of individual nutritional requirements, nutritional status, and genotype (i.e., "individualized nutrition") can be used to prevent, relieve, or cure the chronic disease.

In the following sections, we will further explore nutrigenetics and nutrigenomics in terms of their definition, examples of how each of them works, and conclude with the importance of nutrigenetics and nutrigenomic related to personal nutrition and its impact on personalized medicine.

9.1 Nutrigenetics

Nutrigenetics is the response to dietary components with regard to genetic differences. It describes how the genetic profile has an impact on the response of the body to bioactive food components by influencing their absorption, metabolism, and site of action (Florentino 2007; Mutch et al. 2005).

This term was first used by R.O. Brennan in "Nutrigenetics." As briefly described before, nutrigenetics explores the interaction between gene and diet in terms of how diet or certain nutrients can become a risk factor for monogenetic and polygenetic diseases. Genetic mutations or certain SNPs can result in adverse responses when certain nutrients are consumed (Florentino 2007; Mutch et al. 2005; Ordovas and Mooser 2004). This is representative in several metabolic diseases. We listed a few examples of the relationship between the manifestations of certain SNPs in metabolic diseases and their impact of diets below:

9.1.1 Example 1: Phenylketonuria (PKU)

PKU is a rare inherited disorder caused by different mutations in the phenylalanine hydroxylase gene (PKU) which is responsible for creating the enzyme needed to break down phenylalanine. This ultimately leads to defects in iron absorption and long-chain fatty acid oxidation(Palou 2007). From babies to adults, these patients must maintain a restrictive diet that limits foods that contain proteins or the artificial sweetener and aspartame to prevent the accumulation of phenylalanine(Mutch et al. 2005).

9.1.2 Example 2: Galactosemia

Galactosemia are disorders of galactose metabolism caused by mutations in galactose-1-phosphate uridyltransferase (GALT) gene. These disorders can result in a number of life-threatening complications in infants including feeding problem, hepatocellular damage, bleeding, and sepsis in untreated infants. Treatment available for these patients includes a restriction of galactose intake. This includes a restrictive diet of non-galactose carbohydrates and all-lactose contain foods and products throughout their life span(Berry 2012; Ko et al. 2010).

9.1.3 Example 3: Favism Disease

Favism disease, also called glucose-6-phosphate dehydrogenase deficiency, is a hereditary condition caused by mutations in glucose-6-phosphate dehydrogenase (G6PD) gene. G6PD deficiency is the most common metabolic disorder in the world, estimated to impact 400 million persons throughout the world. Lack of G6PD compromises the body's ability to protect against oxidative stress. These patients can experience an acute hemolytic crisis induced by ingestion of fava beans(Ames et al. 2002; Hedayat et al. 1981; Salvemini et al. 1999).

Nutrigenetics also aims at defining the mechanism by which genes influence nutrients in either of these ways: digestion, absorption, metabolism, distribution, storage, transformation, and excretion by proteins in the form of receptors, carriers, enzymes, hormones, taste perception, and satiation degree. This is not only important to understand the etiology of the disease but to also understand the development of proper treatment as in preventative diets, drug metabolism, and potential side effects of different drugs (Simopoulos 2010a). We have listed a few examples below of the how the imbalance of micronutrients and macronutrients due polymorphisms impacts individual's susceptibility of diseases.

9.1.4 Example 1: Folate in Heart Disease and Cancer

One's metabolism can be largely attributed to polymorphism in the methylene tetrahydrofolatereductase gene (MTHR). This variation is linked to the nutrient folate, as well as affecting one's susceptibility to cancer and heart disease. The two most common and important mutations in the MTHFR gene are C677T and A1298C polymorphisms. C677T polymorphism is a cytosine-to-thymidine substitution resulting in the conversion of alanine to valine. and it is the most common variant that occurs as homozygous T/T in 5–10% of the and as heterozygous C/T genotypes up to 40% general population. A1298C is adenine-to-cytosine substitution resulting in the conversion of an alanine to glutamic acid. The presence of C677T or A1298C mutations is associated with a reduction in MTHFR enzyme activity and impairs folate accumulation. This may lead to increased homocysteine concentration in plasma, a risk factor for venous thromboembolic and ischemic arterial diseases. Another less common variant in the MTHFR gene is Ala222Val. This polymorphism increases the conversion of deoxyuridine monophosphate (dUMP) to deoxythymidine monophosphate (dTMP) and leads to more folate-dependent thymidine biosynthesis and folate deficiency(Stover 2006). This causes an increased risk of spontaneous abortions and decreased fetal viability. Women with such mutation are recommended to consume maternal folate supplements (Stover 2004).

MTHFR gene is also involved in the maintenance of CpG methylation patterns and prevents damaging DNA through strand breaks. Mutations in this gene increase the risk for neural tube defects and some types of cancer(Stover 2006). Changes in the concentration of folate which is the MTHFR substrate and riboflavin considered the MTHFR cofactor should control the activity of the MTHFR gene(Virgili and Perozzi 2008). In these cases, folic acid supplementation is recommended to mitigate the risks or the negative side effects of SNPs in the MTHFR gene(Stover 2006; Virgili and Perozzi 2008).

9.1.5 Example 2: Dietary Cholesterol and Plasma Cholesterol Levels

Serum cholesterol response to dietary cholesterol is genotype-dependent. Male subjects with an Apo E3/4 genotype could lower their low-density lipoprotein (LDL) cholesterol up to 23% by following a low-fat/low-cholesterol diet, while male subjects with an Apo E3/3 genotype lowered their LDL cholesterol up to 14% only and was lowered by 13% in males with Apo E3/2. Furthermore, LDL lowering was twice as great in male subjects when compared to females, when all subjects followed the same diet (Simopoulos 2010a).

9.1.6 Example 3: Omega-6 Fatty Acids and Breast Cancer

Studies suggest based on epidemiology that omega-6 fatty acids increase the risk for certain cancers and omega-3 fatty acids decrease that risk. However, some studies were inconclusive when exploring the correlation between omega-6 fatty acid consumption and risk for breast cancer. It is noteworthy that these inconclusive studies did not take into account the genetic predisposition to metabolize omega-6 fatty acids. Studies that took into account the genetic predisposition found a significant correlation between consumption of omega-6 fatty acids, predisposition to metabolize them, and breast cancer in women(Simopoulos 2010a).

9.1.7 Example 4: Sodium and Blood Pressure

High blood pressure or hypertension is a multifactorial disease that is a result of several factors that could be genetic, nutritional, or environmental; examples of these factors are obesity, sodium, chloride, alcohol, low potassium, low calcium, low omega-3 fatty acid intake, stress, and physical inactivity. However, recent genome-wide association studies (GWAS) identified 13 SNPs for systolic blood pressure and 20 for diastolic blood pressure and 10 for hypertension. Further studies are needed to further analyze the interaction or the effect of these SNPs and different nutrients, so a better understanding of the disease and treatment is developed(Simopoulos 2010a, b).

9.2 Nutrigenomics

Nutrigenomics is the study of the relationship between dietary components and the genome. This study makes it possible to develop a genetic understanding of how common dietary components affect the balance between health and disease by altering the expression and/or structure of an individual's genetic makeup (Kaput and Rodriguez 2004; Mutch et al. 2005). The field of nutrigenomics focuses on the exploration of the effects of different nutrients on the genome, proteome, and metabolome(Ordovas and Mooser 2004; Simopoulos 2010a). Nutrients are broken down into macronutrients and micronutrients. Both types could influence the transcription of a genome, the degree of gene expression, and manifest a heterogeneous response of different gene variants. Nutrigenomics also studies the use of functional genomic tools to analyze how nutritional molecules affect metabolic pathways and homeostatic control. Nutrigenomics requires the usage of transcriptomics, proteomics, and metabolomics technologies to collect and study correlations between food and the human genome. Findings in the field of nutrigenomics promises the possibility for doctors and nutritionists to create personalized nutritional plans for individuals based on their specific genomic makeup(Mutch et al. 2005; Bush et al. 2019).

Several epidemiology studies correlate the intake of certain nutrients with the incidence and severity of chronic diseases, for example, obesity, type two diabetes, cardiovascular disease, and some cancers; while all these diseases are complex and multifactorial, there has been a clear correlation between diet and multiple gene expression in patients(Iacoviello et al. 2008; Jenkins et al. 2003).

A keyway of altering gene expression through nutrients is by transcription factors (TFs). An example of this is how nutrients bind to a class of ligand-activated TFs called peroxisome proliferator-activated receptors (PPARs). This superfamily of TFs influences the expression of genes involved in numerous metabolic processes that take place in the liver, such as fatty acid oxidation, ketogenesis, gluconeogenesis, and cellular proliferation(Afman and Müller 2006).

9.2.1 Example 1: Omega-6 and Omega-3 Fatty Acids

Dietary cholesterol has been found to have a great inhibitory impact on transcribing the gene responsible for β-hydroxy-β-methyl-glutaryl (HMG)-CoA reductase. Dietary polyunsaturated fatty acids (PUFAs) have been found to suppress the production of the fatty acid synthase for lipoproteinemia by suppressing the hepatic mRNA (Simopoulos 2010a).

9.2.2 Example 2: EGCG Consumption and Breast Cancer

It has been found that dietary chemicals should affect the pathways responsible for signal transduction. Green tea contains polyphenol, 11-epigallocatechin-3-gallate (EGCG). The EGCG inhibits tyrosine phosphorylation of Her-2/neu receptor and epidermal growth factor receptor that reduces signaling via the phosphatidylinositol 3-kinase (PI-3)-AKT kinase-NF-kB pathway. The activation of the NF-kB pathway is correlated with certain breast cancers(Nobel et al. 2001; Edwards 2000).

9.2.3 Example 3: Carbohydrates Consumption and Gene Expression Regulation

A study was done on 47 subjects with metabolic syndromes (obesity, hypertension, dyslipidemia, hyperinsulinemia, insulin resistance, and hyperglycemia, predisposes to cardiovascular disease and type two diabetes) who consumed two different diets one characterized with low postprandial insulin response (consuming rye pasta) and another with a high postprandial insulin response (consuming oat-wheat and potato). The study found downregulation of 71 genes in the group that consumed rye pasta diet, and these genes include genes that are linked to insulin signaling and apoptosis, while the other group had upregulated around 62 genes that are related to stress, cytokine-chemokine-mediated immunity, and the interleukin pathway and even though no change in weight was observed among groups, the insulinogenic index improved for the rye-pasta group. This study indicated how diet can have a direct impact on gene expression in patients with metabolic syndromes, regardless of whether the diet incurred a weight loss or not (Simopoulos 2010a).

9.2.4 Example 4: Restricting Caloric Intake and Its Impacts on Gene Expression

Studies found that changes in the daily caloric intake, reduction, from 28 days to 8 weeks had a great impact on regulating gene expression, while some genes were upregulated due to the reduced intake, others were downregulated when compared to their level of expression before the diet. An example of this change is a reduced expression of genes responsible for regulating the production of PUFAs(Simopoulos 2010a).

9.3 Personalized Nutrition

With the growing field of nutrigenomics comes the potential for the application of this field in order to personalize one's diet. Recently, human health research and understanding has shifted from a broad, population-based perspective to a more individualized viewpoint and personal application. Similarly, to the concept of personalized medicine, personalized nutrition contrasts to the traditional, all-encompassing nutrition recommendations that are commonly used and instead focuses on the needs of an individual. Personalized nutrition, or precision nutrition, can be defined as dietary recommendations based on personal habits such as lifestyle and physiological functions and responses to food such as epigenetic changes(Bush et al. 2019).

Genetic research in general is very new and is extremely cutting edge, with mass amounts of potential for personalizing one's medicine and health care. Genetics in general are a huge determining factor when it comes to the outlook of one's health. One's diet similarly can majorly affect one's health, but the degree and manner of this is largely dependent on the person and in turn their personal genetics. Due to this, personalizing one's diet to match their genetics is a very logical step that could revolutionize health care. It is not just the consumer that could be affected by the growing use of nutrigenomics, however. Nutrigenomics additionally could revolutionize the food production industry. For example, the genomes of animals could be used to optimize the tissue that is being used for consumption, as well as helping to learn how to best reduce toxins negatively affecting the food. Additionally, microorganisms could be studied to determine a way to intentionally incorporate them into food products in order to increase their health value and quality(German et al. 2011).

There are a range of nutrients that all humans need a sufficient amount of in their diet in order to remain healthy. This list has already been determined, which means that the current question is how to ensure that everyone reaches the amounts needed. While many people with many different diets get more than enough of these nutrients, some people do not, and this is in many cases genetic. A more complete list of a variety of these nutrients, along with certain genotypes that can negatively affect intake, can be seen in Table 9.1.

Due to one's nutrient intake being largely dependent on one's genotype, determining what genes exactly are responsible and what problems are not genetic is of vital importance. Due to the far-reaching consequences of even just one gene, further genotyping and gene experiments are vital to the future of the field of personalized nutrition.

Several factors, other than genetics, must also be considered when implementing a personal nutrition regime such as environment, lifestyle, taste, culture, and stages of life such as pregnancy or old age. Environment can be described as factor external to the body that still affect nutrition such as sun exposure and vitamin D production. Taste and culture can determine food choice and availability and can have implication in nutrition. Physical activity levels also determine nutritional and caloric requirements and contribute to the effect of lifestyle(Kussmann and Fay 2008).

Table 9.1 Interactions between essential nutrients and gene polymorphisms on clinical outcomes

Nutrient	Gene polymorphism	Effects on nutrient status	Clinical manifestations	Reference
Calcium	Calcium sensing receptor (CASR) A986S	Loss of function for calcium, associated with higher serum calcium, and higher urinary calcium excretion	Association with bone mineral density	Laaksonen et al. (2009)
Selenium	Missense mutation in selenium binding protein 2 (SBP2)	Causes defective selenocysteine insertion sequence (SECIS)-driven selenocysteine incorporation, downregulates expression of selenoproteins	Defective thyroid function	Hesketh (2008)
Iron	Human hemochromatosis protein (HFE) 187C>G or 845G>A	Bothl87C>G or 845G>A associated with iron overload (hemochromatosis)	Iron overload, liver cirrhosis, and cardiomyopathy, especially in diets high in iron	Hulgan et al. (2008)
Folate	5,10-methlenetetrahydrofolate reductase (MTHFR) 677C>T	Causes a 70% reduction in MTHFR activity, hyperhomocysteinemia and reduced plasma folate concentration	Hyperhomocysteinemia is associated with increased risk of coronary heart disease, neural tube defects, occlusive vascular disease and breast cancer. In carriers, sufficient folate dietary intake decreases risk of colorectal cancer, and deficiencies increase risk of colorectal cancer	Ericson et al. (2009), Friso et al. (2002), Hustad et al. (2004), Messika et al. (2010), Simopoulos (2010a)
Sodium	Angiotensin gene (AGT) nucleotide-6 G>A	The A substitutions in AGT affects the interaction between at least one *trans*-acting nuclear factor and its promoter, resulting in increased gene transcription and increased angiotensin protein levels	Carriers of the A allele respond to low sodium diets with reductions in blood pressure; GG genotype is not salt-sensitive	Simopoulos (2010a)

(continued)

Table 9.1 (continued)

Nutrient	Gene polymorphism	Effects on nutrient status	Clinical manifestations	Reference
Vitamin D	Vitamin D binding protein DBP-1 (rs7041, exon 11 T>G) and DBP-2 (rs4588, exon 11C>A)	SNPs for DBP-1 and DBP-2 are inversely related to levels of circulating 25(OH) vit D_3 in premenopausal women	Unclear whether carriers would benefits from dietary supplementation or sun exposure	Sinotte et al. (2009)
Vitamin K	Vitamin K epoxide reductase complex subunit 1 (VKORC1)j – +22557T>C	Associated with vitamin K recycling, vitamin K-dependent clotting factors and Warfarin resistance	Increased risk of arterial vascular disease such as stroke, coronay heart disease, and aortic dissection	Suh et al. (2009)
Vitamin A	β-carotene 15,15-′-monooxygenase (BCMO1) R267S (rsl2934922) andA379V (rs7501331)	Carriers of 267S or 267S+ 379Vhave reduced activity in converting B-carotene to retinal	Increased risk for vitamin A deficiency, when B carotene is the major dietary source	Leung et al. (2009)
Vitamin B12 (cobalamin)	Methionine synthase TCN2 776C>G and 67A>G	Causes hyperhomocysteinemia	Associated with birth defects	Brouns et al. (2008)
Carbohydrates	Beta-2-adrenegic receptors Q27E	Unknown	Higher risk of obesity in female carriers with carbohydrate intake >49% of energy	Martinez et al. (2003)
Omega 3 and 6 fatty acids	Fatty acid desaturase, FADS SNP rs 174537	Lower plasma arachidonic and eicosapentaenoic acids and higher plasma alpha linolenic and linoleic acids in carriers of the minor allele versus noncarriers	The minor allele homozygotes (TT) have lower plasma total cholesterol and LDL-C compared with noncarriers	Tanaka et al. (2009)

To assess the effect of personalized nutrition, Several epidemiology studies provided subjective assessments such as patient motivation or behavior and objective assessments such as data from genomic, epigenomic, proteomic, and microbiome researches. These biochemical "omic" studies show insight into physiological response to changes in diet and can be used to assess the use of personalized nutrition as a therapy(Bush et al. 2019). For example, preliminary studies in epigenetics found that diets lacking necessary levels of folate predispose mice to alterations in DNA methylation patterns and impaired DNA repair(Dauncey 2014). These assessments and parameters of personalized nutrition will aid in clinical interventions and further understanding of how diet can molecularly affect health and lead to a holistic personalized medicine approach for patients.

References

Afman L, Müller M (2006) Nutrigenomics: from molecular nutrition to prevention of disease. J Am Diet Assoc 106(4):569–576. https://doi.org/10.1016/j.jada.2006.01.001

Ames BN, Elson-Schwab I, Silver EA (2002) High-dose vitamin therapy stimulates variant enzymes with decreased coenzyme binding affinity (increased km): relevance to genetic disease and polymorphisms. Am J Clin Nutr 75(4):616–658. https://doi.org/10.1093/ajcn/75.4.616

Berry GT (2012) Galactosemia: when is it a newborn screening emergency? Mol Genet Metab 106(1):7–11. https://doi.org/10.1016/j.ymgme.2012.03.007

Brouns R, Ursem N, Lindemans J, Hop W, Pluijm S, Steegers E, Steegers-Theunissen R (2008) Polymorphisms in genes related to folate and cobalamin metabolism and the associations with complex birth defects. Prenat Diagn 28:485–493. [PubMed: 18435414]

Bull C, Fenech M (2008) Genome-health nutrigenomics and nutrigenetics: nutritional requirements or 'nutriomes' for chromosomal stability and telomere maintenance at the individual level. Proc Nutr Soc 67(2):146–156. https://doi.org/10.1017/s0029665108006988

Bush CL, Blumberg JB, El-Sohemy A et al (2019) Toward the definition of personalized nutrition: a proposal by the American nutrition association. J Am Coll Nutr 39(1):5–15. https://doi.org/10.1080/07315724.2019.1685332

Dauncey MJ (2014) Nutrition, the brain and cognitive decline: insights from epigenetics. Eur J Clin Nutr 68(11):1179–1185. https://doi.org/10.1038/ejcn.2014.173

Edwards P (2000) Regulation of gene expression by SREBP and SCAP. Biochim Biophys Acta Mol Cell Biol Lipids 1529(1–3):103–113. https://doi.org/10.1016/s1388-1981(00)00140-2

Ericson U, Sonestedt E, Ivarsson MI, Gullberg B, Carlson J, Olsson H, Wirfält E (2009) Folate intake methylenetetrahydrofolate reductase polymorphisms and breast cancer risk in women from the Malmö Diet and Cancer cohort. Cancer Epidemiol Biomark Prev 18(4):1101–1110. https://doi.org/10.1158/1055-9965.EPI-08-0401

Florentino RF (2007) Nutrigenomics—opportunities in Asia. In: Karger S (ed) ILSI's first international conference on nutrigenomics: opportunities in Asia, pp 224–241. https://doi.org/10.1159/000107207

Friso S, Choi SW, Girelli D, Mason JB, Dolnikowski GG, Bagley PJ, Olivieri O, Jacques PF, Rosenberg IH, Corrocher R, Selhub J (2002) A common mutation in the 510-methylenetetrahydrofolate reductase gene affects genomic DNA methylation through an interaction with folate status. Proc Natl Acad Sci 99(8):5606–5611. https://doi.org/10.1073/pnas.062066299

German JB, Zivkovic AM, Dallas DC, Smilowitz JT (2011) Nutrigenomics and personalized diets: what will they mean for food? Annu Rev Food Sci Technol 2(1):97–123. https://doi.org/10.1146/annurev.food.102308.124147

Hedayat SH, Farhud DD, Montazami K, Ghadirian P (1981) The pattern of bean consumption, laboratory findings in patients with Favism, G-6-P-D deficient, and a control group. J Trop Pediatr 27(2):110–113. https://doi.org/10.1093/tropej/27.2.110

Hesketh J (2008) Nutrigenomics and selenium: gene expression patterns, physiological targets, and genetics. Annu Rev Nutr 28:157–177. https://doi.org/10.1146/annurev.nutr.28.061807.155446

Hulgan T, Tebas P, Canter JA et al (2008) Hemochromatosis gene polymorphisms, mitochondrial haplogroups, and peripheral lipoatrophy during antiretroviral therapy. J Infect Dis 197(6):858–866. https://doi.org/10.1086/528697

Hustad S, Nedrebø BG, Ueland PM, Schneede J, Vollset SE, Ulvik A, Lien EA (2004) Phenotypic expression of the methylenetetrahydrofolate reductase 677C→T polymorphism and flavin cofactor availability in thyroid dysfunction. Am J Clin Nutr 80(4):1050–1057. https://doi.org/10.1093/ajcn/80.4.1050

Iacoviello L, Santimone I, Latella MC, de Gaetano G, Donati MB (2008) Nutrigenomics: a case for the common soil between cardiovascular disease and cancer. Genes Nutr 3(1):19–24. https://doi.org/10.1007/s12263-008-0079-0

Jenkins DJA, Kendall CWC, Marchie A et al (2003) The Garden of Eden—plant based diets, the genetic drive to conserve cholesterol and its implications for heart disease in the 21st century. Comp. Biochem. Physiol. A Mol. Integr. Physiol. 136(1):141–151. https://doi.org/10.1016/s1095-6433(02)00345-8

Kaput J, Rodriguez RL (2004) Nutritional genomics: the next frontier in the postgenomic era. Physiol Genomics 16(2):166–177. https://doi.org/10.1152/physiolgenomics.00107.2003

Ko D-H, Chang HE, Song SH et al (2010) Molecular and biochemical characterization of the GALT gene in Korean patients with galactose-1-phosphate uridyltransferase deficiency. Clin Chim Acta 411(19–20):1506–1510. https://doi.org/10.1016/j.cca.2010.06.008

Kussmann M, Fay LB (2008) Nutrigenomics and personalized nutrition: science and concept. Pers Med 5(5):447–455. https://doi.org/10.2217/17410541.5.5.447

Laaksonen MML, Outila TA, Kärkkäinen MUM et al (2009) Associations of vitamin D receptor, calcium-sensing receptor and parathyroid hormone gene polymorphisms with calcium homeostasis and peripheral bone density in adult Finns. J Nutrigenet Nutrigenomics 2(2):55–63

Leung WC, Hessel S, Méplan C, Flint J, Oberhauser V, Tourniaire F, Hesketh JE, von Lintig J, Lietz G (2009) Two common single nucleotide polymorphisms in the gene encoding β-carotene 1515′-monoxygenase alter β-carotene metabolism in female volunteers. FASEB J 23(4):1041–1053. https://doi.org/10.1096/fj.08-121962

Martinez JA, Corbalan MS, Sánchez-Villegas A, Forga L, Marti A, Martínez-Gonzalez MA (2003) Obesity risk is associated with carbohydrate intake in women carrying the Gln27Glu beta2-adrenoceptor polymorphism. J Nutr 133:2549–2554

Messika AH, Kaluski DN, Lev E, Iakobishvili Z, Shohat M, Hasdai D, Mager A (2010) Nutrigenetic impact of daily folate intake on plasma homocysteine and folate levels in patients with different methylenetetrahydrofolate reductase genotypes. Eur J Cardiovasc Prev Rehabil 17(6):701–705. https://doi.org/10.1097/HJR.0b013e32833a1cb5

Mutch DM, Wahli W, Williamson G (2005) Nutrigenomics and nutrigenetics: the emerging faces of nutrition. FASEB J. 19(12):1602–1616. https://doi.org/10.1096/fj.05-3911rev

Nobel S, Abrahmsen L, Oppermann U (2001) Metabolic conversion as a pre-receptor control mechanism for lipophilic hormones. Eur J Biochem 268(15):4113–4125. https://doi.org/10.1046/j.1432-1327.2001.02359.x

Ordovas JM, Corella D (2004) Nutritional genomics. Annu Rev Genomics Hum Genet 5(1):71–118. https://doi.org/10.1146/annurev.genom.5.061903.180008

Ordovas JM, Mooser V (2004) Nutrigenomics and nutrigenetics. Curr Opin Lipidol 15(2):101–108. https://doi.org/10.1097/00041433-200404000-00002

Palou A (2007) From nutrigenomics to personalised nutrition. Genes Nutr 2(1):5–7. https://doi.org/10.1007/s12263-007-0022-9

Paoloni-Giacobino A, Grimble R, Pichard C (2003) Genetics and nutrition. Clin Nutr 22(5): 429–435. https://doi.org/10.1016/s0261-5614(03)00064-5

Salvemini F, Franzé A, Iervolino A, Filosa S, Salzano S, Ursini MV (1999) Enhanced glutathione levels and oxidoresistance mediated by increased glucose-6-phosphate dehydrogenase expression. J Biol Chem 274(5):2750–2757. https://doi.org/10.1074/jbc.274.5.2750

Simopoulos AP (2010a) Nutrigenetics/nutrigenomics. Annu Rev Public Health 31(1):53–68. https://doi.org/10.1146/annurev.publhealth.031809.130844

Simopoulos AP (2010b) Genetic variants in the metabolism of omega-6 and omega-3 fatty acids: their role in the determination of nutritional requirements and chronic disease risk. Exp Biol Med 235(7):785–795. https://doi.org/10.1258/ebm.2010.009298

Sinotte M, Diorio C, Bérubé S, Pollak M, Brisson J (2009) Genetic polymorphisms of the vitamin D binding protein and plasma concentrations of 25-hydroxyvitamin D in premenopausal women. Am J Clin Nutr 89(2):634–640. https://doi.org/10.3945/ajcn.2008.26445

Stover PJ (2004) Nutritional genomics. Physiol Genomics 16(2):161–165. https://doi.org/10.1152/physiolgenomics.00204.2003

Stover PJ (2006) Influence of human genetic variation on nutritional requirements. Am J Clin Nutr 83(2):436S–442S. https://doi.org/10.1093/ajcn/83.2.436s

Suh J-W, Baek S-H, Park J-S, Kang H-J, Chae I-H, Choi D-J, Park H-J, Kim P-J, Seung K-B, Kim H-S (2009) Vitamin K epoxide reductase complex subunit 1 gene polymorphism is associated with atherothrombotic complication after drug-eluting stent implantation: 2-Center prospective cohort study. Am Heart J 157(5):908–912. https://doi.org/10.1016/j.ahj.2009.02.001

Tanaka T, Shen J, Abecasis GR, Kisialiou A, Ordovas JM et al (2009) Genome-wide association study of plasma polyunsaturated fatty acids in the InCHIANTI Study. PLoS Genet 5:e1000338

Virgili F, Perozzi G (2008) How does nutrigenomics impact human health? IUBMB Life 60(5): 341–344. https://doi.org/10.1002/iub.85

Chapter 10
Nutrigenomics and Development: Childhood Obesity Susceptibility Genes and their Impact on Dietary Behavior and Nutrient Intake

Emily Burch

Abstract Childhood obesity is considered a serious global public health issue. An increasing number of children are affected by obesity each year. Genetics are known to significantly contribute to the development of childhood obesity. Many obesity susceptibility genes have been identified and evaluated in terms of their relationship to nutrient intake and dietary behavior. Several of these obesity susceptibility genes, including FTO, MC4R, and BDNF, have been shown to increase intake of calorie-dense food and promote excessive weight gain. Evaluating the gene-diet interactions is a key element in understanding the mechanisms of childhood obesity. Increased understanding of the impact that genes have on dietary behavior will aid in the development of personalized approaches in the prevention and treatment of childhood obesity.

10.1 Introduction

Childhood obesity is a significant problem worldwide. The number of overweight or obese children is continually increasing. In 1990, it was estimated that there were about 32 million overweight or obese children (0–5 years of age). This number increased to a staggering 41 million globally in 2016. A large majority of overweight or obese children are living in developing countries. The rate of increase in developing countries is about 30% higher than that of developed countries (WHO 2017). Obesity can have profound effects on child's physical health, social well-being, psychological health, and self-esteem. Childhood obesity is also associated with poor academic performance and lower quality of life. Children who are overweight or obese have an increased likelihood of being obese in adulthood. They are also at a

E. Burch (✉)
Institute of Global Health and Human Ecology (IGHHE), The American University in Cairo (AUC), New Cairo, Egypt
e-mail: burche@aucegypt.edu

M. Salama (ed.), *Nutrigenomics and the Brain*, Nutritional Neurosciences,
https://doi.org/10.1007/978-981-16-9205-5_10

higher risk of developing a variety of health problems including diabetes and cardiovascular disease at a younger age (Bhadoria et al. 2015).

The mechanisms that contribute to childhood obesity are still not completely understood at this time. A range of factors that are commonly believed to attribute to the development of childhood obesity include, but are not limited to, environmental, lifestyle, cultural factors, and biological (Bhadoria et al. 2015). The significance of genetics to the development of childhood obesity has been the subject of many recent studies. It is suggested that genetic factors account for 40–90% of variations in body mass index (Duran-Gonzalez et al. 2011).

The term "obesogenic environment" is defined as the sum of influences that promote obesity in individuals or populations. Currently, there is an increase in availability of high-calorie foods and lifestyles have become more sedentary. Both of these elements contribute to the growing obesity epidemic. However, it is noted that not all individuals become overweight or obese when exposed to the same environment. These noted variations may result from obesity susceptibility genes and their interactions with the obesogenic environment (Garver et al. 2013). Identifying and understanding susceptibility genes will further our understanding of the mechanisms behind childhood obesity and aid in the development of future preventative strategies and treatments.

The purpose of this article is to review the origins, mechanisms, and interactions of obesity susceptibility genes and to evaluate the impact they have on dietary choices and behaviors in childhood obesity. Furthering our understanding of this complex topic may provide valuable insight for future therapies and may highlight the need for more personalized approaches to combating the global obesity epidemic.

10.2 Measures of Obesity

Body mass index (BMI) is a tool that is often used to define overweight and obesity. BMI uses an individual's weight and height to measure body fat. The calculation for BMI is weight in kilograms divided by height in meters squared (kg/m2). Measurements of BMI in children also account for age and gender by the use of percentile charts. In the United States, a child is considered overweight if their BMI is greater than or equal to the 85th percentile, and a child is considered obese if their BMI is greater than or equal to the 95th percentile. Many studies use BMI as an indicator for childhood obesity; however, the definition parameters may very slightly between studies.

The term energy balance refers to the relationship between energy consumption, energy expenditure, and energy storage. A positive energy balance occurs when the body stores excess energy primarily in the form of fat. A negative energy balance occurs when the body draws on those stores of energy to provide energy for work (Garver et al. 2013). Excess weight gain is usually a result of positive energy balance. Research suggests that any genetic or environmental factor that influences body weight must involve the basic components of energy balance over a period of time (Hill et al. 2012).

10.3 Nutrigenomics

Gene diet interactions are considered to be significant contributors to many chronic nutrition-related diseases, such as obesity and type 2 diabetes (Flowers et al. 2016). The discovery of this relationship promoted the development of a scientific discipline referred to as nutrigenomics. Nutrigenomics is the study of the interaction between nutrition and genes, especially with regard to the prevention and treatment of disease. The definitive role of epigenetic and environmental interactions and their relationship to obesity remain controversial due to noted difficulties in discriminating between the cause and effect of the DNA modifications. It is still unclear whether epigenetic modifications cause obesity or if obesity causes epigenetic modifications (Martin et al. 2011; Garver et al. 2013). However, the importance of this relationship has been widely studied especially with respect to obesity susceptibility genes.

10.4 Factors of Obesogenic Environment

The obesogenic environment consists of a complex interplay of contributing factors that influence behavior and effect dietary choice, physical activity, or metabolism responsible for maintaining energy balance (Patrick et al. 2004). In simple terms, it is any environment that encourages a person to eat unhealthy and to not participate in physical activity. Many recent studies suggest that sedentary behavior and reduced overall physical activity alongside shorter sleep duration promote the overconsumption of dietary macronutrients, particularly fats and refined carbohydrates (Garver et al. 2013). Studies suggest that increased consumption of a high-fat diet, primarily a diet high in saturated fatty acids, has been strongly associated with increased adiposity in overweight and obese children. Consistent with these results, it is also indicated that obesity susceptibility genes tend to preferentially interact with saturated fatty acids, but not monounsaturated fatty acids or polyunsaturated fatty acids, to promote weight gain (Razquin et al. 2009).

10.5 Types of Childhood Obesity

10.5.1 Syndromic Obesity

Syndromic obesity is characterized by severe obesity that is accompanied by multisystem disorders like intellectual disability, retinal degeneration, sensorineural deafness, organ-specific abnormalities, and dysmorphic features (Maltese et al. 2018). Rare monogenic forms of syndromic obesity are represented by approximately 30 unidentified susceptibility genes. Some examples of syndromic obesity include Prader-Willi, Bardet-Biedl, Alstrom, Carpenter, Rubstein-Taybi, and Cohen syndromes (Garver et al. 2013). In general, children with syndromic obesity have extreme adiposity, physical dysmorphology, and intellectual disabilities. Neuroendocrine abnormalities are commonly associated with some of syndromic obesity. These neuroendocrine abnormalities are suggested to be responsible for having adverse effects on the hypothalamus.

The hypothalamus serves as the brain appetite center. One of its many roles is to regulate energy balance through energy consumption and energy expenditure (Farooqi and O'Rahilly 2005). It maintains energy homeostasis by receiving information about the body's current nutritional status and energy stores through neural and hormonal signals (Bell et al. 2005; Vetter et al. 2010; Fonseca et al. 2017). Due to the adverse effects on the hypothalamus, children with syndromic obesity are usually characterized with severe hyperphagia, meaning increased appetite and intake, and diminished satiety (Bray 1992; Garver et al. 2013). Disorders related to syndromic obesity are often complex, involving several overlapping and unidentified loci that are suggested to contribute to alterations in the regulation of energy balance (Garver et al. 2013).

10.5.2 Non-Syndromic Obesity

Non-syndromic obesity is characterized by weight gain and the absence of other clinical symptoms. Non-syndromic obesity results from single gene mutations. There are a number of susceptibility genes that are considered to be responsible for this rare monogenic form of obesity. Examples of these include brain-derived neurotrophic factor (BDNF), leptin (LEP), leptin receptor (LEPR), melanocortin-4 receptor (MC4R), neurotrophic tyrosine kinase receptor type 2 (NTRK2), prohormone convertase 1 (PCSK1), proopiomelanocortin (POMC), and single-minded homolog 1 (SIM1). The susceptibility genes in non-syndromic obesity encode for enzymes and receptors that have a physiologic role in the leptin/melanocortin system. The leptin/melanocortin pathway is located within the hypothalamus and is an important pathway controlling food intake and regulating body weight (Fonseca et al. 2017).

10.5.3 Common Obesity

It has been well characterized that the obesity epidemic is not caused by single gene disorders. It is the result of a complex genetic background contributing to obesity. Common obesity is comprised of many genetic susceptibility variants that have minor effects on body weight (Huvenne et al. 2016). The effect of genes on the pathogenesis of obesity is stronger when combined with other genes and the environment (Fonseca et al. 2017). It is interesting to note that among the obesity susceptibility genes that have been identified using genome-wide association studies (GWAS), different variants of the same gene (FTO, MC4R, and BDNF) may also be responsible for rare non-syndromic forms of childhood obesity (Garver et al. 2013). Common obesity susceptibility genes interact with the environment to promote positive energy balance and weight gain. This highlights the fact that obesity is a complex genetic and metabolic disorder (Garver et al. 2013).

10.6 Susceptibility Genes

10.6.1 Fat Mass and Obesity-Associated Protein

The fat mass and obesity-associated gene (FTO) was first identified as a predictor of BMI (Silva et al. 2018). Research supports that many SNPs in the FTO gene are not only associated with an increase in BMI but also potentially related to other factors including glucose metabolism, increased triglycerides, reduced HDL cholesterol, increased weight, eating behavior, and waist circumference (Fredriksson et al. 2008; Rivas et al. 2018). The rs9939609 SNP, which is located within the first intron of the FTO gene, has been studied extensively and is found to have a strong association with obesity in children and adolescents as well as adults. The mechanism by which FTO polymorphisms influence human obesity remains unclear.

The relationship between weight gain and an FTO gene-diet interaction was first established in rats when reduction in the amounts of mRNA FTO was observed during a period of fasting, leading to increased food consumption, while feeding had the opposite effect (Tung et al. 2010). The results of this study indicate that appetite is influenced by that expression and the functional amount of FTO in the hypothalamus (Garver et al. 2013). Many studies have evaluated FTO gene variants and their relationship with energy consumption. These studies noted that children with FTO gene variants were associated with increased and preferential consumption of high-calorie foods, particularly to foods that were enriched with saturated fatty acids (Cecil et al. 2008; Bauer et al. 2009).

A cross-sectional study was conducted on Chilean children with the aim of assessing the association between the rs9939609s SNP in the FTO gene and variations in eating behavior patterns in Chilean children. In children, some investigators found that carriers of the susceptibility allele A show increased energy intake in relation to the carriers of the T allele. It is widely known that energy intake is a measure characterized by wide variability; therefore, the study of eating behavior traits appears to be a better approach. Thus, the aim of this study was to assess the association between the genetic variant rs9939609 in the FTO gene and homeostatic and non-homeostatic eating behavior patterns in Chilean children (Rivas et al. 2018). This study revealed some associations between the FTO SNP and eating behavior traits that may predispose children to excessive energy intake and promote weight gain. Other studies have suggested that weight gain in children with the FTO gene variant resulted from abnormal eating behavior characterized by a loss of control eating episodes of foods high in fat (Rivas et al. 2018). An earlier study conducted on children and adolescents also evaluated the relationship between the rs9939609 gene variant in FTO and eating behavior. This study found that children with FTO rs9939609 obesity-risk alleles reported more frequent loss of control eating episodes. The children and adolescents also selected foods with higher fat content at a buffet meal. Loss of control eating and the increased frequency of selecting high-calorie foods are suggested mechanisms through which variant FTO alleles lead to excess body weight (Tanofsky-Kraff et al. 2009).

10.6.2 Brain-Derived Neurotrophic Factor

Brain-derived neurotrophic factor (BDNF) is a member of the neurotrophin family of growth factors. It has been identified as a key component in the hypothalamic pathway. BDNF plays an important role in the regulation of energy homeostasis and food intake through hypothalamic signaling (Roth et al. 2013). BDNF and its receptor tropomyosin-related kinase B (trkB) are extensively expressed in areas of the hypothalamus that are associated with feeding and metabolism (Roth et al. 2013).

Humans with mutations in the BDNF gene or trkB receptor signal transduction pathway exhibit severe obesity (Roth et al. 2013). It has been evaluated that obese patients with variants in BDNF may present with hyperphagia and insulin resistance. However, a study conducted to evaluate BDNF and its relation to leptin in obese children, prior to and following weight loss, had confounding results. This study found no significant relationship between BDNF and insulin resistance. The results showed increased concentrations of BDNF in obese children when compared to children of normal weight. BDNF was noted to be significantly correlated to adiposity. These results could suggest a relationship between BDNF and fat mass, possibly through leptin signaling in the brain (Roth et al. 2013). BDNF has been shown to increase food intake in mice; however, there are limited studies on the influence of BDNF on eating behavior in humans (Rios 2010).

10.6.3 Melanocortin 4 Receptor

The MC4R gene is a member of the seven transmembrane G-coupled protein receptor family. The relationship between obesity and MC4R had been extensively evaluated in rats. However, MC4R was not linked to extreme childhood obesity until the discovery of a frameshift mutation (Yeo et al. 1998). It was later suggested that individuals possessing MC4R gene variants were at an increased risk of obesity. A study conducted in 2003 compared eating behavior of 20 obese carriers of MC4R gene variants and 120 non-carriers with similar characteristics. This study also included one normal weight carrier of an MC4R gene mutation. In this study, all carriers of an MC4R mutation met criteria for a diagnosis of binge-eating disorder, and only 14.2 percent of non-carriers met the same criteria. The results of this study suggest that binge eating is a major phenotype responsible for excess weight gain in patients with MC4R gene variations (Branson et al. 2003). The results of this study were supported by a GWAS aimed at identifying chromosomal regions contributing to increased energy consumption. This study was conducted on 1030 Hispanic children and found that MC4R gene variants play a key role in the regulation of body weight. MC4R gene variations were associated with increased energy consumption and decreased energy expenditure in children (Cai et al. 2006).

10.7 Discussion

A key factor in combating the global childhood obesity epidemic is to understand the basic mechanisms for how gene variants predispose weight gain. Examples observed in this study are the FTO and MC4R gene variants that lead to increased energy intake. Children with either gene variant showed increased preference for calorie-dense foods. Children with MC4R gene variations also showed decreased propensity for energy expenditure. Identifying the root cause of childhood obesity is vital for combating this epidemic. The results of the studies in this review highlight the need for personalized approaches in the prevention and treatment of childhood obesity.

Many childhood obesity intervention programs traditionally focus on generalized population guidelines. New insights from nutrigenomic studies serve an important role in both prevention and treatment of childhood obesity (Papoutsakis and Dedoussis 2007). Currently, there are novel therapeutic interventions that are being developed for certain genetic forms of obesity (Serra-Juhé et al. 2019). These therapeutic interventions highlight the importance of incorporating more personalized approaches in the treatment of childhood obesity. Further advancements in treatment and prevention will be made possible by enhancing our understanding of the mechanisms of childhood obesity and by using targeted nutritional and medicinal therapies (Garver et al. 2013).

10.8 Conclusion

The current epidemic of childhood obesity is represented by a wide variety of genetic variations in both monogenic and polygenic forms. These genetic variations and their interaction with the obesogenic environment promote childhood obesity. Increasing our understanding of the mechanisms that contribute to childhood obesity should be a priority in combating this epidemic. Further evaluation of gene-diet interaction is needed as well as studies evaluating the relationship between susceptibility genes and the etiology and pathology of nutrition-related diseases. By increasing our understanding in this area of research we will potentially be able to develop personalized nutritional or medicinal therapy that will better address the current obesity problem.

By submitting this exam/quiz/assignment, I affirm that I have followed AUC's Code of Academic Ethics and the work submitted is my own. I have not consulted unauthorized resources or materials nor collaborated with other individuals unless allowed.

References

Bauer F, Elbers CC, Adan RA, Loos RJ, Onland-Moret NC, Grobbee DE et al (2009) Obesity genes identified in genome-wide association studies are associated with adiposity measures and potentially with gene-specific food preference. Am J Clin Nutr 90(4):951–959. https://doi.org/10.3945/ajcn.2009.27781

Bell CG, Walley AJ, Froguel P (2005) The genetics of human obesity. Nat Rev Genet 6(3): 221–234. https://doi.org/10.1038/nrg1556

Bhadoria A, Sahoo K, Sahoo B, Choudhury A, Sufi N, Kumar R (2015) Childhood obesity: causes and consequences. J Fam Med Prim Care 4(2):187. https://doi.org/10.4103/2249-4863.154628

Branson R, Potoczna N, Kral JG, Lentes K-U, Hoehe MR, Horber FF (2003) Binge eating as a phenotype of melanocortin 4 receptor gene mutations. N Engl J Med 349(18):1782–1782. https://doi.org/10.1056/nejm200310303491825

Bray GA (1992) Chapter 24 Genetic, hypothalamic and endocrine features of clinical and experimental obesity. In: Progress in brain research the human hypothalamus in health and disease, proceedings of the 17th international Summer School of Brain Research, held at the auditorium of the University of Amsterdam, pp 333–341. https://doi.org/10.1016/s0079-6123(08)64583-5

Cai G, Cole SA, Butte N, Bacino C, Diego V, Tan K et al (2006) A quantitative trait locus on chromosome 18q for physical activity and dietary intake in Hispanic children. Obesity 14(9): 1596–1604. https://doi.org/10.1038/oby.2006.184

Cecil JE, Tavendale R, Watt P, Hetherington MM, Palmer CN (2008) An obesity-associated FTOGene variant and increased energy intake in children. N Engl J Med 359(24):2558–2566. https://doi.org/10.1056/nejmoa0803839

Duran-Gonzalez J, Ortiz I, Gonzales E, Ruiz N, Ortiz M, Gonzalez A et al (2011) Association study of candidate gene polymorphisms and obesity in a young Mexican-American population from South Texas. Arch Med Res 42(6):523–531. https://doi.org/10.1016/j.arcmed.2011.10.010

Farooqi IS, O'Rahilly S (2005) Monogenic obesity in humans. Annu Rev Med 56(1):443–458. https://doi.org/10.1146/annurev.med.56.062904.144924

Flowers E, Froelicher ES, Aouizerat BE (2016) Gene-environment interactions in cardiovascular disease. Eur J Cardiovasc Nurs. https://doi.org/10.1177/1474515112437591

Fonseca ACPD, Mastronardi C, Johar A, Arcos-Burgos M, Paz-Filho G (2017) Genetics of non-syndromic childhood obesity and the use of high-throughput DNA sequencing technologies. J Diabetes Complications 31(10):1549–1561. https://doi.org/10.1016/j.jdiacomp.2017.04.026

Fredriksson R, Maria H, Olszewski PK, Stephansson O, Jacobsson JA, Olszewska AM et al (2008) The obesity gene, FTO, is of ancient origin, up-regulated during food deprivation and expressed in neurons of feeding-related nuclei of the brain. Endocrinology 149(5):2062–2071. https://doi.org/10.1210/en.2007-1457

Garver WS, Newman SB, Gonzales-Pacheco DM, Castillo JJ, Jelinek D, Heidenreich RA, Orlando RA (2013) The genetics of childhood obesity and interaction with dietary macronutrients. Genes Nutr 8(3):271–287. https://doi.org/10.1007/s12263-013-0339-5

Hill JO, Wyatt HR, Peters JC (2012) Energy balance and obesity. Circulation 126(1):126–132. https://doi.org/10.1161/CIRCULATIONAHA.111.087213

Huvenne H, Dubern B, Clément K, Poitou C (2016) Rare genetic forms of obesity: clinical approach and current treatments in 2016. Obes Facts 9(3):158–173. https://doi.org/10.1159/000445061

Maltese PE, Iarossi G, Ziccardi L, Colombo L, Buzzonetti L, Crinò A et al (2018) A next generation sequencing custom gene panel as first line diagnostic tool for atypical cases of syndromic obesity: application in a case of Alström syndrome. Eur J Med Genet 61(2):79–83. https://doi.org/10.1016/j.ejmg.2017.10.016

Martin DI, Cropley JE, Suter CM (2011) Epigenetics in disease: leader or follower? Epigenetics 6(7):843–848. https://doi.org/10.4161/epi.6.7.16498

Papoutsakis C, Dedoussis GV (2007) Gene–diet interactions in childhood obesity: paucity of evidence as the epidemic of childhood obesity continues to rise. Pers Med 4(2):133–146. https://doi.org/10.2217/17410541.4.2.133

Patrick K, Norman GJ, Calfas KJ, Sallis JF, Zabinski MF, Rupp J, Cella J (2004) Diet, physical activity, and sedentary behaviors as risk factors for overweight in adolescence. Arch Pediatr Adolesc Med 158(4):385. https://doi.org/10.1001/archpedi.158.4.385

Razquin C, Martinez JA, Martinez-Gonzalez MA, Bes-Rastrollo M, Fernández-Crehuet J, Marti A (2009) A 3-year intervention with a Mediterranean diet modified the association between the rs9939609 gene variant in FTO and body weight changes. Int J Obes (Lond) 34(2):266–272. https://doi.org/10.1038/ijo.2009.233

Rios M (2010) New insights into the mechanisms underlying the effects of BDNF on eating behavior. Neuropsychopharmacology 36(1):368–369. https://doi.org/10.1038/npp.2010.139

Rivas AMO, Santos JL, Valladares MA, Cameron J, Goldfield G (2018) Association of the FTO fat mass and obesity–associated gene rs9939609 polymorphism with rewarding value of food and eating behavior in Chilean children. Nutrition 54:105–110. https://doi.org/10.1016/j.nut.2018.03.001

Roth CL, Elfers C, Gebhardt U, Müller HL, Reinehr T (2013) Brain-derived neurotrophic factor and its relation to leptin in obese children before and after weight loss. Metabolism 62(2):226–234. https://doi.org/10.1016/j.metabol.2012.08.001

Serra-Juhé C, Martos-Moreno GÁ, Bou de Pieri F, Flores R, Chowen JA, Pérez-Jurado LA, Argente J (2019) Heterozygous rare genetic variants in non-syndromic early-onset obesity. Int J Obes 44(4):830–841. https://doi.org/10.1038/s41366-019-0357-5

Silva TERD, Andrade NL, Cunha DDO, Leão-Cordeiro JAB, Vilanova-Costa CAST, Silva AMTC (2018) The FTO rs9939609 polymorphism and obesity risk in teens: evidence-based meta-analysis. Obes Res Clin Pract 12(5):432–437. https://doi.org/10.1016/j.orcp.2018.08.001

Tanofsky-Kraff M, McDuffie JR, Yanovski SZ, Kozlosky M, Schvey NA, Shomaker LB, Salaita C, Yanovski JA (2009) Laboratory assessment of the food intake of children and adolescents with loss of control eating. Am J Clin Nutr 89(3):738–745. https://doi.org/10.3945/ajcn.2008.26886

Tung Y-CL, Ayuso E, Shan X, Bosch F, O'Rahilly S, Coll AP, Yeo GS (2010) Hypothalamic-specific manipulation of FTO, the ortholog of the human obesity gene FTO, affects food intake in rats. PLoS One 5(1). https://doi.org/10.1371/journal.pone.0008771

Vetter ML, Faulconbridge LF, Webb VL, Wadden TA (2010) Behavioral and pharmacologic therapies for obesity. Nat Rev Endocrinol 6(10):578–588. https://doi.org/10.1038/nrendo.2010.121

WHO (2017) Childhood overweight and obesity. Retrieved from https://www.who.int/dietphysicalactivity/childhood/en/

Yeo GS, Farooqi IS, Aminian S, Halsall DJ, Stanhope RG, Orahilly S (1998) A frameshift mutation in MC4R associated with dominantly inherited human obesity. Nat Genet 20(2):111–112. https://doi.org/10.1038/2404

Chapter 11
The Effects of Early Childhood Malnutrition on Neurodevelopment

Emily Burch

Abstract It is estimated that undernutrition affected more than 191 million children under the age of 5 in 2019. Early childhood malnutrition impacts health and increases a child's vulnerability for cognitive impairments later in life. Poor nutrition in the first 1000 days of a child's life can lead to stunted growth, which is associated with impaired cognitive ability and reduced school and work performance. This review examined the relationship between early childhood malnutrition and various elements related to neurodevelopment, including intelligence, cognitive functioning, and school performance. This study found that early childhood malnutrition can significantly adversely affect neurodevelopment, and those adverse effects can be observed even into adulthood.

11.1 Introduction

The period between conception and the first 2 years of life, commonly referred to as the first 1000 days, is an important time for neurodevelopment in children (Cusick and Georgieff 2016). Neurodevelopment occurs over a person's entire life span, but the first 1000 days are considered especially critical because of the rapid and complex development that occurs. In this time period, the foundations of optimal growth, health and neurodevelopment are established (Schwarzenberg and Georgieff 2018; Turner and Honikman 2016). Healthy neurodevelopment is a maturational process, resulting in systematic acquisition of motor, perceptual, cognitive, language, psychological, and self-regulation skills (Burger et al. 2020). If a child is not in an optimal state of health during this time period, it can leave them especially vulnerable to suboptimal growth and development.

There are many risk factors that can have a detrimental effect on fetal and child development. These risk factors may be biological, environmental, or conceptual.

E. Burch (✉)
Institute of Global Health and Human Ecology (IGHHE), The American University in Cairo (AUC), New Cairo, Egypt
e-mail: burche@aucegypt.edu

M. Salama (ed.), *Nutrigenomics and the Brain*, Nutritional Neurosciences,
https://doi.org/10.1007/978-981-16-9205-5_11

Examples of biological factors include birth complications and suboptimal childhood nutrition. Environmental factors may include physical surroundings or geographical conditions, and contextual factors may include poor mental health, low education, poverty, and lack of quality health care (Daelmans et al. 2015). There are a number of adverse outcomes that can result from a child not reaching their full developmental potential. Poor academic achievements, economic dependency, and adult onset of noncommunicable diseases are just a few examples of these potential outcomes (Daelmans et al. 2015; Schwarzenberg and Georgieff 2018).

Many studies have evaluated the relationship between nutrition, cognitive function, and neurodevelopment. Studies suggest that chronic protein-energy malnutrition early in life can result in cognitive impairments and can slow the rate of development of cognitive processes (Kar et al. 2008). These cognitive impairments have been noted even into adulthood. This review article will evaluate the impact of malnutrition on neurodevelopment. The effects of malnutrition will be examined on variety of aspects of neurodevelopment including intelligence, cognitive function, and school performance.

11.2 Malnutrition

Malnutrition can be defined as a state in which a deficiency, excess, or imbalance of energy, protein, and/or other nutrients results in measurable adverse effects on the body. The term malnutrition can refer to both undernutrition and overnutrition. Undernutrition is the inadequate intake of nutrients, and overnutrition is the excessive intake of nutrients. UNICEF estimated that 194.7 million children were affected by malnutrition in the form of either stunting or wasting in 2020.

In 1959, the term "protein calorie malnutrition" was coined to embrace two extreme forms of childhood undernutrition, kwashiorkor and marasmus (Jelliffe 1959). Marasmus presents as wasting of muscle tone and adipose tissue resulting from chronic semistarvation or inadequate energy intake (Chopra and Sharma 1992). Marasmus is generally defined as a reduction in body weight below 60% of the normal expected weight for age. Kwashiorkor results from severe protein malnutrition and is characterized by a loss in muscle tone with preservation of adipose tissue and is accompanied by hypoalbuminemia and edema (Chopra and Sharma 1992).

The effects of malnutrition can vary depending on the type of malnutrition present. Studies show that chronic protein energy malnutrition can have adverse effects on ongoing development of higher cognitive processes during childhood (Kar et al. 2008). Micronutrient deficiencies are also noted to impact neural development. Research shows that vitamins and minerals play important roles on various physiological processes in the brain and influence cognitive function (Lam and Lawlis 2017). A few examples of this include folate, iron, and DHA. Folate deficiency during pregnancy has been associated with congenital abnormalities in the fetus, including neural tube defects (Greenberg et al. 2011). Folate and vitamin B12 are required for biological methylation and DNA synthesis. Folate and vitamin B12 deficiencies are associated with increased risk of neural tube defects among children cognitive impairment, depression, Alzheimer's disease, and stroke among adults (Troen 2012).

11.3 Measures of Malnutrition

Undernutrition manifests in four broad forms including stunting, wasting, underweight, and micronutrient deficiencies. Wasting is defined as low weight for height and often indicates recent and severe weight loss. Wasting is a strong predictor of mortality in children under the age of 5 (UNICEF 2007). A micronutrient deficiency is the lack of vitamins and/or minerals that are essential for body functions such as producing enzymes, hormones, and other substances needed for growth and development. It is estimated that at least half of children worldwide under the age of 5 suffer from vitamin and mineral deficiencies (UNICEF 2019).

Stunting can be defined as a height deficit or impaired growth in children, under the age of 5. Stunting can result from poor nutrition, repeated infection, and/or inadequate psychosocial stimulation (Akombi et al. 2017). Stunting is an indicator of long-term chronic undernutrition. According to the World Health Organization (WHO), children are defined as stunted if their Z-scores representing height-for-age are greater than two standard deviations below the WHO child growth standard median (The WHO 2016). In 2017, it is estimated that 24.6% of children under 5 years of age, in low- and middle-income countries, met the criteria for stunting. Stunting in early life can result in impaired growth and adverse functional effects in children. Some adverse effects could include poor cognition and educational performance, low adult wages, lost productivity and, when accompanied by excessive weight gain later in childhood, an increased risk of nutrition-related chronic diseases in adult life. Prevention of stunting among children under the age of 5 years is listed as one of the World Health Organizations' (WHO) target goals for sustainable development.

11.4 The Effects of Malnutrition

11.4.1 Intelligence Quotient

The Oxford Dictionary defines intelligence as the ability to acquire knowledge and skills. A number of studies have examined the relationship between early malnutrition and intelligence by the use of a set of tests referred to as the intelligence quotient (IQ). The Barbados Nutrition Study is a 45-yearlong longitudinal study that evaluates the adverse effects of malnutrition in early childhood. The malnourished subjects in this study were born at a normal weight but experienced an episode of moderate to severe malnutrition, requiring hospitalization, during their first year of life. These individuals have been followed through childhood, adolescence, and now adulthood, along with a case-control comparison group from the same community.

In 1987, the long-term effects of early kwashiorkor and marasmus were examined in terms of intellectual performance based on a cohort from the Barbados Nutrition Study (Galler et al. 1987a, b). Criteria for an individual to be included in the malnourished group was a diagnosis during the first year of life of kwashiorkor, or marasmus, a birth weight of greater than 5 pounds, no evidence of pre- or postnatal complications, and no history of convulsions, head injury, or loss of consciousness during childhood. Kwashiorkor was characterized by edema, hair, and skin changes. Marasmus was defined as weight loss below 75% of expected weight for age and the absence of edema. A low Apgar score was used to confirm that there were no pre- or postnatal complications (Galler et al. 1987a, b). This study included 53 participants who met the criteria for kwashiorkor, 55 for marasmus, and 65 healthy controls who ranged from 11 to 18 years of age. This study then evaluated full-scale IQ scores for each of the three groups. This study found that the children of the kwashiorkor and marasmus groups had IQ scores that were similar to each other but significantly less than those of the healthy comparison group (Galler et al. 1987a, b). The largest difference noted between the two groups occurred within the age group of 14.6 to 16.5 years of age. This study indicates that cognitive deficits are associated with early malnutrition, and they are not alleviated in adolescence (Galler et al. 1987a, b). Similar findings were noted in a longitudinal study of 190 international adoptees that were followed from birth to age 23. This study evaluated the IQ of international adoptees, who met criteria for malnutrition, at the age of 7, 14, and 23. This data was then compared to that of a normal group of infants, and the results indicated that early malnutrition negatively affected the adoptees' IQ. The results of this study noted a greater impact on IQ in middle childhood and a lesser effect on IQ in adulthood (Schoenmaker et al. 2015).

In 2014, a study was conducted to evaluate impaired IQ and academic skills in adults who experienced early malnutrition. This study was conducted on a group of individuals who have all been followed since childhood in the Barbados Nutrition Study. Assessments were conducted on IQ and academic skills on 76 previously malnourished adults and 59 controls with a mean age of 38.4 and 38.1, respectively (Waber et al. 2013). Adult IQ was assessed using the Wechsler Abbreviated Scale of Intelligence (WASI). The results of the IQ test were analyzed, and 26.3% of the previously malnourished group were identified to have impaired IQ compared to only 3% in the control group. This supports the idea that early malnutrition can have a lifetime impact on IQ. The findings of this study also suggest that cognitive and behavioral development remained adversely affected in the previously malnourished subjects despite catchup in their physical growth by adolescence (Waber et al. 2013).

11.4.2 Cognitive Functioning

Learning, thinking, reasoning, remembering, problem-solving, decision-making, and attention are all included in cognitive function (Fisher et al. 2019). The term cognitive impairment refers to a temporary or permanent loss of mental functions. Many studies have evaluated the relationship between cognitive impairments and malnutrition. A study conducted in India investigated the effect of stunting, as a result of protein energy malnutrition (PEM), on the nature of cognitive impairments and the rate of cognitive development (Kar et al. 2008). This study used the NIMHANS neuropsychological battery for a comprehensive neuropsychological assessment of children 5–15 years of age. This neuropsychological assessment evaluates motor speed, attention, executive functions, visuospatial relationships, comprehension, learning, and memory (Kar et al. 2004). The children were assessed for malnutrition based on their anthropometrics. Children were included in the malnourished group if they met the criteria for stunting alone or a combination of stunting and wasting in reference to the national center of health statistics (NCHS) standards of growth and development. The malnourished group of children were matched with adequately nourished children from the same school and of the same age and grade level. Children were excluded from the study if they met criteria for mental retardation (Kar et al. 2008).

Following statistical analysis, this study showed that the malnourished group differed significantly from the adequately nourished group on a number of tests including phonemic fluency, design fluency, selective attention, visuospatial functions, verbal comprehension and learning, and memory showing performance. The two groups did not differ on the finger tapping tests, and the malnourished group did not have any difficulty with respect to expressive speech (Kar et al. 2008). The results of this study show that malnourished children, in comparison to adequately nourished children, had poor performance on almost all of the neuropsychological tests except that of motor speed.

The findings of this study, indicating that malnourished children showed poor performance on tests of higher cognitive functions, are also supported by another study conducted in India. In this study, malnourished children showed poor performance on a number of novel tasks. These tasks included tests of executive functions like working memory and spatial locations. It was also noted that performance on the finger dexterity test for fine motor coordination was not significantly affected in the group of undernourished children (Agarwal et al. 1995). This further supports the idea that malnutrition early in life can negatively impact cognitive performance. The findings related to motor speed and fine motor coordination are confounded by a study conducted in 1987 that was conducted on a cohort from the Barbados Nutrition Study. This study used the Purdue Pegboard Test and compared scores on children, aged 11–18, with a history of marasmus to children with a history of kwashiorkor and a control group. The results of this study showed significant deficits in three of the four categories examined. One category, the assembly test, the kwashiorkor group had more extensive deficits even when compared to the marasmus group. This study suggests that early malnutrition can have adverse effects on fine motor skills and nervous system function even up to 18 years of age (Galler et al. 1987a, b).

A study conducted in Ghana worked to evaluate the relationship between dietary micronutrients and cognitive performance tests. This study evaluated the dietary intake of micronutrients in school-aged children to assess for any deficiencies. The study then utilized the Raven's Colored Progressive Matrices (RCPM) to assess the cognitive level of the schoolchild. This study found that mean cognition test scores did not differ by age, and adequate or inadequate intake of iron, zinc, vitamin B6, vitamin B12, and vitamin A intake. However, one exception was noted, folate. The children with adequate folate intake had a higher mean RCPM test score than those with inadequate intake. This implies that folate intake was related to the cognition test score (Annan et al. 2019). The relationship between micronutrient intake and cognition is still being evaluated due to noted inconsistencies. An example of this is the relationship between cognition and vitamin B12 intake. A study conducted in the United States found no association between dietary vitamin B12 and cognitive outcomes (Boeke et al. 2013). However, studies in Kenya and Iran found that higher dietary vitamin B12 was associated with improved cognitive outcomes intake in school-aged children (Gewa et al. 2009; Ahmadi et al. 2009).

11.4.3 School Performance

A study conducted in India aimed to evaluate the relationship between the prevalence of malnutrition and scholastic performance in primary and secondary school children. This study utilized anthropometric data including BMI, weight for age Z-scores and height for age Z scores of students enrolled in two private schools in the rural Bangalore district. The anthropometric data was then compared to English and mathematic exam scores. The results showed that decreasing weight for age was significantly associated with decreasing scores in mathematics. Height for age was significantly associated with first language scores, as height for age decreased so did first language scores (Rashmi et al. 2015). Similar findings were noted in a study conducted in South Africa. This study noted a positive association between "health status" and educational achievement for both English and Mathematics (Themane et al. 2003). Weight for age and BMI were strongly associated with both English and mathematical performance. Height for age was only positively associated with mathematical performance. There are a number of noted additional factors that can influence a child's academic performance including learning environment, socio-economic status, race, and gender (Themane et al. 2003). The study conducted by Thermane et al. noted that the relationships identified were hardly influenced by age and gender, suggesting that "health status" may be a significant contributor.

An assessment on performance of national high school entrance exams was conducted on a group of 11-year-old Barbadian children (Galler et al. 1990). These children were part of the Barbados Nutrition Study cohort. This study found that a history of early malnutrition was associated with impaired performance on national 11-plus examinations. The scores of children with a history of malnutrition were found to be significantly lower than that of the control group. No significant differences were noted between the marasmus and kwashiorkor groups. This study notes that poor performance is strongly associated with behavioral and cognitive deficits. This study also worked to correct for the possible effects of home environmental conditions. After adjusting for the current home environments of the children, the relationship between exam scores and classroom behavior was unchanged. Findings of this study suggest early malnutrition may result in restrictions in educational opportunities and could potentially have lifelong implications.

11.5 Discussion

The results of this study highlight different aspects of the life-span cognitive burden of early malnutrition. This study focuses on early childhood malnutrition either in the first year of life, through the Barbados nutrition study or early on in development, represented by the presence of stunting. Impairments in cognitive functioning, IQ, and school performance have been noted. Although there are other factors believed to contribute to these impairments, like age, sex, socioeconomic status, and environment, many of the studies discussed above attempted to account for these factors to assess the true burden of malnutrition, while other studies had noted no significant differences between the demographics of sex and age. It is impossible to know if malnutrition is the greatest contributor to these cognitive impairments, but evidence from a number of studies suggest that it is a prominent factor.

The effects of supplementation of protein and micronutrients in early childhood have also been evaluated. However, there are conflicting results in both micronutrient supplementation and macronutrient supplementation. This may be a result of timing of interventions in relation to the presence of malnutrition or baseline nutritional status at the time of intervention. A study conducted in Guatemala found that protein supplementation during infancy had beneficial effects on cognitive outcomes and schooling at age 40 (Pollitt et al. 1995). Another study conducted in New Jersey found no improvements in school performance following 8 months of vitamin B12 supplementation. Further studies need to be conducted to evaluate the benefits of supplementation (Perlman et al. 2010). However, the prevention of early malnutrition is ideal for promoting optimal growth and development early in life. The importance of preventing malnutrition has been described by many organizations including the WHO. In 2018, the WHO stated that the eradication of extreme poverty and hunger is one of their Millennium Developmental Goals (The WHO 2016).

Expanding research within this area of study to include other disciplines will help to further expand our understanding of the mechanisms of cognitive impairments related to early malnutrition. This increased understanding could aid in the development of preventative and therapeutic treatments. In 2016, a study was conducted to assess DNA methylation signatures associated with impairments in attention and cognition in relation to early childhood malnutrition. This study was conducted on subjects from the Barbados Nutrition Study. The results of this study identified more than 100 genomic loci that showed DNA methylation changes after early childhood malnutrition. This study concluded a strong link between protein malnutrition in early childhood and epigenetic dysregulation associated with deficits in attention and cognition in adults. This study strengthens the associations between early malnutrition and cognitive impairments later in life (Peter et al. 2016). Future epigenetic studies present a great opportunity to further evaluate the effects of early malnutrition.

11.6 Conclusion

There are many studies that indicate that early childhood malnutrition adversely effects neurodevelopment. Early childhood malnutrition has been found to have negative effects on the outcomes of IQ tests, cognitive assessments, and evaluations of school performance. The effects of early childhood malnutrition have been noted in all stages of life from childhood to adolescence and even into adulthood. There are a substantial number of studies showing associations between early childhood malnutrition and cognitive impairments. However, studies evaluating the effects of micro- and macronutrient supplementation have noted inconsistencies. The prevention of malnutrition is seemingly the best way to prevent cognitive impairments later on in life. Further studies must be conducted from a variety of disciplines to aid in our understanding of the full effects of early childhood malnutrition and neurocognitive development.

By submitting this exam/quiz/assignment, I affirm that I have followed AUC's Code of Academic Ethics and the work submitted is my own. I have not consulted unauthorized resources or materials nor collaborated with other individuals unless allowed.

References

Agarwal K, Agarwal D, Upadhyay S (1995) Impact of chronic undernutrition on higher mental functions in Indian boys aged 10–12 years. Acta Paediatr 84(12):1357–1361. https://doi.org/10.1111/j.1651-2227.1995.tb13568.x

Ahmadi A, Sohrabi Z, Eftekhari M (2009) Evaluating the relationship between breakfast pattern and short-term memory in junior high school girls. Pak J Biol Sci 12(9):742–745. https://doi.org/10.3923/pjbs.2009.742.745

Akombi BJ, Agho KE, Hall JJ, Merom D, Astell-Burt T, Renzaho AM (2017) Stunting and severe stunting among children under-5 years in Nigeria: a multilevel analysis. BMC Pediatr 17(1). https://doi.org/10.1186/s12887-016-0770-z

Annan RA, Apprey C, Asamoah-Boakye O, Okonogi S, Yamauchi T, Sakurai T (2019) The relationship between dietary micronutrients intake and cognition test performance among school-aged children in government-owned primary schools in Kumasi metropolis, Ghana. Food Sci Nutr 7(9):3042–3051. https://doi.org/10.1002/fsn3.1162

Boeke CE, Gillman MW, Hughes MD, Rifas-Shiman SL, Villamor E, Oken E (2013) Choline intake during pregnancy and child cognition at age 7 years. Am J Epidemiol 177(12): 1338–1347. https://doi.org/10.1093/aje/kws395

Burger M, Hoosain M, Einspieler C, Unger M, Niehaus D (2020) Maternal perinatal mental health and infant and toddler neurodevelopment—evidence from low and middle-income countries. A systematic review. J Affect Disord 268:158–172. https://doi.org/10.1016/j.jad.2020.03.023

Chopra J, Sharma A (1992) Protein energy malnutrition and the nervous system. J Neurol Sci 110(1–2):8–20. https://doi.org/10.1016/0022-510x(92)90003-4

Cusick SE, Georgieff MK (2016) The role of nutrition in brain development: the golden opportunity of the "first 1000 days". J Pediatr 175:16–21. https://doi.org/10.1016/j.jpeds.2016.05.013

Daelmans B, Black MM, Lombardi J, Lucas J, Richter L, Silver K et al (2015) Effective interventions and strategies for improving early child development. BMJ. https://doi.org/10.1136/bmj.h4029

Fisher GG, Chacon M, Chaffee DS (2019) Theories of cognitive aging and work. In: Work across the lifespan, pp 17–45. https://doi.org/10.1016/b978-0-12-812756-8.00002-5

Galler JR, Ramsey FC, Forde V, Salt P, Archer E (1987a) Long-term effects of early kwashiorkor compared with marasmus. II Intellectual performance. J Pediatr Gastroenterol Nutr 6(6): 847–854. https://doi.org/10.1097/00005176-198711000-00005

Galler JR, Ramsey FC, Salt P, Archer E (1987b) Long-term effects of early kwashiorkor compared with marasmus. III Fine motor skills. J Pediatr Gastroenterol Nutr 6(6):855–859. https://doi.org/10.1097/00005176-198711000-00006

Galler JR, Ramsey FC, Morley DS, Archer E, Salt P (1990) The long-term effects of early kwashiorkor compared with marasmus. IV Performance on the National High School Entrance Examination. Pediatr Res 28(3):235–239. https://doi.org/10.1203/00006450-199009000-00018

Gewa CA, Weiss RE, Bwibo NO, Whaley S, Sigman M, Murphy SP, Harrison G, Neumann CG (2008) Dietary micronutrients are associated with higher cognitive function gains among primary school children in rural Kenya. Br J Nutr 101(09):1378. https://doi.org/10.1017/s0007114508066804

Greenberg JA, Bell SJ, Guan Y, Yu YH (2011) Folic acid supplementation and pregnancy: more than just neural tube defect prevention. Rev Obstet Gynecol 4(2):52–59

Jelliffe D (1959) Protein-calorie malnutrition in tropical preschool children. J Pediatr 54(2): 227–256. https://doi.org/10.1016/s0022-3476(59)80067-6

Kar BR, Rao SL, Chandramouli BA, Thennarasu K (2004) NIMHANS neuropsychological battery for children. PsycTESTS Dataset. https://doi.org/10.1037/t67515-000

Kar BR, Rao SL, Chandramouli BA (2008) Cognitive development in children with chronic protein energy malnutrition. Behav Brain Funct 4(1):31. https://doi.org/10.1186/1744-9081-4-31

Lam LF, Lawlis TR (2017) Feeding the brain—the effects of micronutrient interventions on cognitive performance among school-aged children: a systematic review of randomized controlled trials. Clin Nutr 36(4):1007–1014. https://doi.org/10.1016/j.clnu.2016.06.013

Perlman AI, Worobey J, Maillet JO, Touger-Decker R, Hom DL, Smith JK (2010) Multivitamin/mineral supplementation does not affect standardized assessment of academic performance in elementary school children. J Am Diet Assoc 110(7):1089–1093. https://doi.org/10.1016/j.jada.2010.04.004

Peter CJ, Fischer LK, Kundakovic M, Garg P, Jakovcevski M, Dincer A, Amaral AC, Ginns EI, Galdzicka M, Bryce CP, Ratner C, Waber DP, Mokler D, Medford G, Champagne FA, Rosene DL, McGaughy JA, Sharp AJ, Galler JR, Akbarian S (2016) DNA methylation signatures of early childhood malnutrition associated with impairments in attention and cognition. Biol Psychiatry 80(10):765–774. https://doi.org/10.1016/j.biopsych.2016.03.2100

Pollitt E, Gorman KS, Engle PL, Rivera JA, Martorell R (1995) Nutrition in early life and the fulfillment of intellectual potential. J Nutr 125(Suppl. 4):S1111–S1118

Rashmi M, Shweta B, Fathima F, Agrawal T, Shah M, Sequeira R (2015) Prevalence of malnutrition and relationship with scholastic performance among primary and secondary school children in two select private schools in Bangalore Rural District (India). Indian J Community Med 40(2):97. https://doi.org/10.4103/0970-0218.153871

Schoenmaker C, Juffer F, Ijzendoorn MHV, Dries LVD, Linting M, Voort AVD, Bakermans-Kranenburg MJ (2015) Cognitive and health-related outcomes after exposure to early malnutrition: the Leiden longitudinal study of international adoptees. Child Youth Serv Rev 48:80–86. https://doi.org/10.1016/j.childyouth.2014.12.010

Schwarzenberg SJ, Georgieff MK (2018) Advocacy for improving nutrition in the first 1000 days to support childhood development and adult health. Pediatrics 141(2). https://doi.org/10.1542/peds.2017-3716

The WHO (2016) Child Growth Standards. Retrieved from https://www.who.int/childgrowth/standards/en/

Themane M, Monyeki K, Nthangeni M, Kemper H, Twisk J (2003) The relationship between health (malnutrition) and educational achievements (Maths and English) in the rural children of South Africa. Int J Educ Dev 23(6):637–643. https://doi.org/10.1016/s0738-0593(03)00063-4

Troen AM (2012) Folate and vitamin B12: function and importance in cognitive development. In: Meeting Micronutrient requirements for health and development, Nestlé Nutrition Institute Workshop Series, pp 161–171. https://doi.org/10.1159/000337684

Turner RE, Honikman S (2016) Maternal mental health and the first 1 000 days. S Afr Med J 106(12):1164. https://doi.org/10.7196/samj.2017.v106i12.12129

UNICEF (2007) Stunting, wasting and overweight. Retrieved from https://www.unicef.org/progressforchildren/2007n6/index_41505.htm

UNICEF (2019) The state of the world's children 2019: children, food and nutrition. Retrieved from https://data.unicef.org/resources/state-of-the-worlds-children-2019/

Waber DP, Bryce CP, Girard JM, Zichlin M, Fitzmaurice GM, Galler JR (2013) Impaired IQ and academic skills in adults who experienced moderate to severe infantile malnutrition: a 40-year study. Nutr Neurosci 17(2):58–64. https://doi.org/10.1179/1476830513y.0000000061

Chapter 12
The Emerging Role of Vitamin D Deficiency as a Risk Factor of Parkinson's Disease

Mohamed El-Gamal, Jihan Azar, and Refaat Hegazi

Abstract Parkinson's disease (PD) is the second most common neurodegenerative disorder of the elderly affecting around 1% of individuals older than 65 years of age. PD poses a major health and economic burden, due to its chronic and progressive nature and the lack of available effective medications to stop or even slow its progression. Majority of PD cases is not linked to known genetic mutations and could be related to environmental factors and an unhealthy lifestyle including deficiency of micronutrients like vitamin D. In the current work, we reviewed the literature to understand the role of vitamin D in the brain and the association between vitamin D insufficiency and deficiency and risk of developing PD. We focus on the molecular mechanisms explaining the increased PD risk concentrating on the nutrigenomic effects of vitamin D. Due to the fact that the prevalence of PD in Egypt is higher than the international one, vitamin D deficiency and some other modifiable risk factors for PD among the Egyptian were discussed. The potential therapeutic role of vitamin D for managing PD was further reviewed. Finally, we could draw the conclusion that vitamin D deficiency may be associated with an increase in the risk and progression of PD. More research is warranted to elucidate the role of vitamin D as preventive and therapeutic options for patients with PD.

Keywords Vitamin D · Parkinson's disease · Inflammation · Diabetes mellitus · Insulin resistance

---

M. El-Gamal (✉)
Toxicology Department, Faculty of Medicine, Mansoura University, Mansoura, Egypt

Medical Experimental Research Center (MERC), Mansoura University, Mansoura, Egypt
e-mail: doctorelgamal@mans.edu.eg

J. Azar
The American University in Cairo, New Cairo, Egypt

R. Hegazi
Abbott Nutrition Research and Development, Columbus, OH, USA

M. Salama (ed.), *Nutrigenomics and the Brain*, Nutritional Neurosciences,
https://doi.org/10.1007/978-981-16-9205-5_12

12.1 Epidemiology of Parkinson's Disease (PD)

Parkinson's disease (PD) is considered to be the second most common neurodegenerative disorder in the elderly which affects approximately 1% of the population older than 60 years (Hirtz et al. 2007). The prevalence of PD in the USA is expected to increase by 2.25-folds, and in Europe, by 1.83-fold between 2010 and 2050 (Bach et al. 2011). Furthermore, the crude prevalence of PD in the USA is expected to increase from 0.401% in 2005 to 0.535% by 2040, and the number of PD patients will reach 700, 000 individuals (Rossi et al. 2018).

As the disease progresses, both the motor and non-motor manifestations worsen. Until now, there is no effective therapy available to halt or even to slow down the disease progression. Indeed, symptomatic treatment is the only available therapy and is associated with many side effects (Oertel 2017; Toulouse and Sullivan 2008). L-dopa is one of the gold standard treatments for PD. However, it induces many side effects including dyskinesia in approximately 30–80% of patients with PD (Phillips et al. 2016). Additionally, it may even enhance progression of the disease by increasing the fraction of free dopamine induced-oxidative stress (Peritore et al. 2012).

PD exerts a major economic burden due to direct and indirect factors. It was estimated that the annual cost of the PD care is around 20,000 US dollars. Direct cost is mainly driven by healthcare services, medications, deep brain stimulation, and physiotherapy. Indirect cost could be driven by reduction of work performance and hours of patients with PD, absentees from work, and care of patients by family members (Bovolenta et al. 2017; Dowding et al. 2006; Kowal et al. 2013; Martinez-Martin et al. 2019). It has been reported that reduction of PD progression to half could reduce such cost to about third (Johnson et al. 2013).

The lack of specific etiology of PD has led investigators to study the association between PD and plethora of genetic and environmental factors that could play a role in its pathogenesis. One of these factors is vitamin D.

12.2 Association Between Low Level of Vitamin D and Risk of PD

Epidemiological studies indicate a positive association of vitamin D deficiency and risk of PD. A prospective study was conducted in Finland between 1978 and 1980 (Mini-Finland health survey), which included 3173 individuals not suffering from PD. Around 29 years later, 50 individuals developed PD. High level of vitamin D seems to show a protective effect against PD as concluded from the low relative risk (RR) between the highest and lowest quartile (RR: 0.33, 95%CI 0.14–0.8; Knekt et al. 2010). Contrarily, another prospective study contradicted these findings not reporting a protective effect of vitamin D against the risk of developing PD. Levels of vitamin D were assessed among 12,762 individuals, and the samples were collected from 1990 to 1992. At the end of 2008, 67 PD patients were identified. No association was observed between the level of vitamin D and risk of PD, as the hazard ratios (HRs) were (1.05: 95% CI: 0.58–1.90) when vitamin D levels are between 20 and 30 ng/mL and (1.14: 95% CI: 0.59–2.23) when vitamin D levels equal or higher than 30 ng/mL compared to individuals having lower levels than 20 ng/mL (Shrestha et al. 2016). These contradictory results between these studies could be attributed to differences in sample sizes of the individual studies, vitamin D assays, and PD diagnostic validation, dictating the need for more studies.

Systematic review of the literature including human observational studies assessing the relationship between vitamin D and PD risk shows an inverse relationship between vitamin D levels and risk of developing PD. One review included 7 studies where 1008 patients and 4536 controls were assessed. Both vitamin D insufficiency and deficiency increased the risk of PD ((odds ratio (OR): 1.5, 95% CI: 1.1–2.0)) and (OR: 2.2, 95% CI: 1.5–3.4), respectively (Lv et al. 2014). Consistently, another systemic review investigating the relationship between vitamin D and PD included both human and rodents studies. The review also supported the hypothesis that vitamin D can have both protective and symptomatic effects against PD (Rimmelzwaan et al. 2016).

A meta-analysis of 8 studies including 5690 PD patients and 21,251 controls that were published until March 2015 showed an increased risk of PD among individuals with vitamin D insufficiency and deficiency (OR: 1.29, 95% CI: 1.10–1.51) and (OR: 2.08, 95% CI: 1.63–2.65), respectively. A more recent meta-analysis of 20 studies published until January 2018 including 2866 PD patients and 2734 individuals reported that reduced vitamin D levels were associated with an increase in the risk of PD. Vitamin D deficiency and insufficiency were associated with PD risk (OR: 2.08, 95% CI: 1.35–3.19) and (OR: 1.73, 95% CI: 1.48–2.03), respectively. Moreover, reduced vitamin D level increases the severity of PD ($r = -0.55$, 95% CI -0.73, -0.29; Luo et al. 2018). Notably, the authors reported high heterogeneity depending on the assay methods. Furthermore, vitamin D supplementation reduced PD risk (OR: 0.62, 95% CI: 0.35–0.90: Shen and Ji 2015).

In conclusion, it could be suggested that low vitamin D levels is associated with increased risk of PD. In the next section, we will review potential molecular and non-molecular factors that could explain such association.

12.3 How Could Vitamin D Alter the Risk of Developing PD on the Molecular Level?

12.3.1 Vitamin D Synthesis and Its Role in the Brain

Vitamin D can be found in food, or it can be synthesized by the human skin when exposed to sunlight. Our skin first synthesizes the precursor for vitamin D, 25-(OH)D. This is further hydroxylated in the liver and then in the kidney to form

1,25 dihydroxyvitamin D (1,25-(OH)2D3), or calcitriol, which is the bioactive form of vitamin D. These hydroxylations are carried out by the enzyme 1a-hydroxylase. Calcitriol activates vitamin D receptor (VDR) and regulates gene transcription (Rimmelzwaan et al. 2016). While VDR is a nuclear receptor, 1a-hydroxylase is found in the cytoplasm (Lương and Nguyễn 2012). Both vitamin D receptor and 1a-hydroxylase are found in neurons and glial cells. The active form of vitamin D, 1,25-(OH)2D, induces its neuroprotective action in several ways. First, it inhibits nitric oxide synthase which produces the free radical, nitric oxide. It also induces the activity of γ-glutamyl transpeptidase, which synthesizes glutathione. Glutathione is an important antioxidant and free radical neutralizer. In addition to its neuroprotective actions, 1,25-(OH)2D induces the production of growth factors like nerve growth factor (NGF), glial derived neurotrophic factor (GDNF), and neurotrophin 3 (NT3) (Rimmelzwaan et al. 2016).

When activated, microglia, the immune cells of the nervous system, transform into phagocytes, identical to macrophages. Microglia are implicated in the pathogenesis of PD. Activated microglia lead to perpetuation of inflammation and neuronal cell death which progresses to dopaminergic neurodegeneration in PD. The substantia nigra has the highest number of microglia, rendering it the neuronal organ most vulnerable to inflammation and neurodegeneration. The substantia nigra contains mostly dopaminergic neurons and is hypothesized to be the location of pathology for PDMicroglial activation negatively affects the structure and function of dopaminergic neurons. Dopaminergic neurons in the substantia nigra also possess lower levels of glutathione, a powerful antioxidant, which reduces their antioxidant activity and therefore significantly increase susceptibility to microglia even more (Yan et al. 2014).

The activation of microglia can be mediated by interferon gamma (IFN-γ), inducible nitric oxide synthase (iNOS), interleukin 1 beta (IL-1β), and tumor necrosis factor alpha (TNF-α).

1a-hydroxylase is found in the substantia nigra neurons as well as glial cells and neurons of the hypothalamus (Lương and Nguyễn 2012). Vitamin D deficiency is associated with tyrosine hydroxylase gene suppression. Tyrosine hydroxylase is an essential enzyme needed for dopamine synthesis and neurotrophic factors (Zhou et al. 2019).

In PD, iNOS expression is increased, which aggravates dopaminergic neuronal death. The chronic inflammatory process, stimulated by the activation of pro-inflammatory cytokines which activate microglia, is thought to be leading pathology to the progression of PD (Yan et al. 2014). iNOS expression was attenuated in cortical neurons when exposed to vitamin D treatment. The INOS gene has VDR response element on it. As VDR is one of the receptors for vitamin D, this highlights the potential direct action of vitamin D on iNOS (Dursun et al. 2014).

12.3.2 Association of VDR Polymorphism and PD

Observational studies comparing VDR genetic variation rs10735810 (FokI) C allele in PD patients and controls have shown mixed results. While some studies showed an increased frequency of the C allele of VDR among patients with PD, others noticed an association of the CC allele and mild forms of PD in Japanese patients, which could be explained population variation. Another study showed that the VDR BsmI genotype in Koreans was associated with PD, while the VDR gene increased the susceptibility to PD in Caucasians (Lương and Nguyễn 2012). A study on Hungarian patients showed an association between an increase in the FokI C allele and PD as compared to controls, and other genetic variants did not show any associations (Török et al. 2013). In contrast, a study of Taiwanese population showed no association of genetic variants of the VDR with PD (Lin et al. 2014). Another study in Han Chinese population showed an association between VDR FokI T/C polymorphism and sporadic PD (Han et al. 2012). Other polymorphisms of the VDR gene include ApaI, BsmI, and TaqI. A meta-analysis found no association of these polymorphisms and PD susceptibility, while FokI polymorphism showed an increased risk in the Asian populations specifically (Gao et al. 2010). TNF-a production is induced by microglial activation (Kuno et al. 2005). Calcitriol was found to inhibit the expression of TNF in microglial cells in two ways: increasing IL-10 expression and forming a complex of LITAF (a transcription factor of TNF) and VDR, which prevents the LITAF from binding to the promoter of the TNF gene (He et al. 2017). VDR is known to affect many diseases by regulating gene transcription (Butler et al. 2011). Vitamin D is known for its anti-inflammatory effect in neurodegenerative diseases. Its effect on calcium homeostasis is thought to play a major protective role in the pathogenesis of PD. This could be explained by the fact that dopaminergic neurons in the substantia nigra are vulnerable to the effect of sustained opening of L-type calcium channels (Butler et al. 2011). More studies are needed to confirm the relationship between VDR polymorphism and risk of PD (Rimmelzwaan et al. 2016).

12.3.3 Molecular Pathways of the Association Between Vitamin D and PD

12.3.3.1 MHCII Complex

Major histocompatibility complex (MHC) is an important locus in human DNA which contains many polymorphic genes. They encode for cell surface proteins that are important for the adaptive immune system. Some genes in this region confer to an increased risk of PD. Specifically, increased levels of MHC class II expression were found in monocytes of PD patients.

Astrocytes increase the levels of glial fibrillary acidic protein (GFAP) and glutathione peroxidase to protect dopaminergic neurons. In PD, astrocytes increase the expression of GFAP. Astrocytes also develop type 2 helper T-cell immune responses that suppress IL-12 expression and increase the expression of MHC and stimulatory cytokines. All of which leads to an inflammatory response (Yan et al. 2014). In response to toxic factors, astrocytes release toxic factor which damage the dopaminergic cell bodies which later lead to degeneration of neurons in the striatum (Yan et al. 2014).

Interestingly, 1,25-(OH)2 D3 (calcitriol) is known to suppress MHC class II antigen expression in monocytes and macrophages (Lương and Nguyễn 2012). This inhibitory effect of vitamin D on MHC expression could partly explain the potential protective effect of vitamin D in PD.

12.3.3.2 Cytochrome P450

Cytochrome P450 is a group of enzymes responsible for the metabolism of endogenous as well as exogenous compounds. They can also be found in the brain and are known to be polymorphic and therefore may contribute to diseases, either by exhibiting increased or decreased activity. CYP2D6*4 allele was found to be more prevalent among patients with PD as compared to controls. Interestingly, CYP2D6 can also act as 25-hydroxylase, activating vitamin D3 (formed by the skin) conversion to 25-(OH)D. Deficiency of the 25-hydroxylase enzymes is associated with vitamin D deficiency (Lương and Nguyễn 2012).

Moreover, CYP2D6 polymorphism is also associated with PD. According to ethnicities, CYP2D6 was found to be absent in less than 1% of the Asian population, and in up to 10% of Caucasians. Finding a significant association between this polymorphism and PD would require a large study group, and stratification according to ethnicities.

Of significance is that CYP2D and PD loci are located on chromosome 22. Deletion of chromosome 22q11 resulted in lower serum calcium, bone mineral density, and parathyroid hormone levels (Lương and Nguyễn 2012). CYP2D6 is a potential 25-hydroxylase, converting vitamin D3 to 25OHD. The lack of this enzyme will result in vitamin D deficiency (Lương and Nguyễn 2012).

12.3.3.3 Heme Oxygenase-1 (HO-1)

HO-1 is an inducible cytoprotective enzyme which is expressed in response to oxidative stress. It is responsible for the catabolism of heme to biliverdin and subsequently bilirubin alongside the production of carbon monoxide (CO) (Gozzelino et al. 2010). Low dose of CO has been found to exert anti-inflammatory effects in animal models of inflammatory bowel disease (Hegazi et al. 2005).

While in normal brains, HO-1 is found in low levels. HO-1 was overexpressed in astrocytes of PD especially in substantia nigra and Lewy bodies in dopaminergic neurons. 1,25-(OH) D3/calcitriol could exert its protective effects in PD via HO-1 dependent mechanism. Consistently, calcitriol has been shown to delay HO-1 immunoreactivity after cerebral ischemia (Lương and Nguyễn 2012).

12.3.3.4 Poly(ADP-Ribose) Polymerase-1 (PARP-1)

PARP-1 is a protein acting on the nucleus of the cell, and in response to stress, it can either induce neuronal death or survival (Lương and Nguyễn 2012). The role of PARP-1 in stress response is explained in detail in a review by (Xin Luo and Lee Kraus 2012). MPTP is a well-known neurotoxin that causes parkinsonian symptoms. PARP gene lacking mice have shown to be spared from the effects of MPTP. PARP-1 was also overexpressed in the dopaminergic neurons in the substantia nigra of PD patients. High levels of vitamin D suppress PARP-1 expression, in a dose-dependent manner. Vitamin D is hypothesized to induce this anti-inflammatory effect by inhibiting microglial activation.

12.3.3.5 Neurotrophic Factors (NTFs)

NTFs are important proteins for the survival of neurons (Brockmann et al. 2016). NTFs can promote neuronal regeneration or protect them from insult. NTFs include brain-derived neurotrophic factors (BDNF), glial-derived neurotrophic factor (GDNF), mesencephalic-astrocyte-derived neurotrophic factor (MANF), and cerebral dopamine neurotropic factor (CDNF). Their respective receptors are found in the striatum and the substantia nigra. In PD, expression of NTFs is reduced. In Koreans, CDNF single-nucleotide polymorphism (rs7094179) increased the susceptibility to PD. In the Chinese Han population, an allele of BDNF was a risk factor for PD. Calcitriol acts on the expression of neurotrophic receptors and expression of GDNF. By increasing GDNF and restoring tyrosine hydroxylase expression in the substantia nigra and striatum, calcitriol protects against dopamine loss (Lương and Nguyễn 2012). GDNF is an important neuroprotective agent for dopamine neurons in the midbrain. GDNF administration alleviated symptoms in PD patients and primate models. In vitro, calcitriol increased glutathione, and its precursor γ-glutamyl transpeptidase and neurotrophin-3 as well (Smith et al. 2006). Decreased levels of neurotrophins in the nigrostriatal region of postmortem brain of PD patients were evident in previous studies. These neurotrophins included BDNF. A suggested mechanism for that is microglial activation which subsequently resulted in the death of dopaminergic neurons (Nagatsu and Sawada 2005). Activated monocytes secrete bioactive BDNF during inflammatory processes to aid in neuronal survival (Brockmann et al. 2016). Increased BDNF serum levels were associated with disease duration and severe motor impairment in PD patients. BDNF also correlated positively with inflammatory markers of neurodegenerative diseases (Brockmann et al. 2016).

12.3.3.6 Sp1 Transcription Factor

Sp1 transcription factor is a DNA-binding protein. In response to oxidative stress in neurons, it gets acetylated. Sp1 family is important in the expression of dopamine transporter gene (Lương and Nguyễn 2012). Sp1 sites act synergistically with vitamin D-responsive elements to induce CYP24 (25-OH-D 24-hydroxylase) production which is important for the metabolism of 1,25OH-D (Lương and Nguyễn 2012). A previous study showed that Sp1 inhibition was seen to reduce monoamine oxidase B activity which resulted in neuroprotective effects (Yao et al. 2018).

12.3.4 Nonmolecular Mechanism of the Association Between Vitamin D and PD

Vitamin D, as one of the fat-soluble vitamins, is stored in adipose tissue, and it alters the inflammatory response of adipocytes (Stevens 2021). Hypertrophy of the adipose tissues associated with obesity results in insufficient blood supply, hypoxia, macrophages infiltration, release of pro-inflammatory cytokines (IL6 & 8, MCP1, TNF-alpha and resistin), and altering adipokines secretion. These changes are associated with insulin resistance (de Luca and Olefsky 2008; Heilbronn and Campbell 2008). It was consistently shown in recent literature that DM increases the risk of PD and its clinical progression. This could be attributed to mitochondrial dysfunction, oxidative stress, neuroinflammation, impaired protein hemostasis. Insulin resistance and poor glycemic control worse PD progress (Hassan et al. 2020). The potential beneficial effect of vitamin D in PD could be explained by its anti-inflammatory effects and inhibition of insulin resistance.

Vitamin D decreases insulin resistance by affecting the release of adipokines as it increases adiponectin and inhibits leptin release. Furthermore, vitamin D plays a critical physiological role as a powerful anti-inflammatory molecule by inhibiting P_{38}MAP kinase and NF-kB signaling pathways, reducing the expression of pro-inflammatory factor genes (IL1-beta, IL-8, and TNF-alpha), (Koszowska et al. 2014; Szymczak-Pajor and Śliwińska 2019). It's totally accepted among the scientific community that insulin resistance is observed during developing diabetes mellitus (DM: Reusch 2002; Taylor et al. 1994) and vitamin D deficiency increases the risk of developing DM through losing the protective roles, as reviewed in details by Berridge (2017). The prevalence of vitamin D deficiency among Kenyan patients with DM was 38.4% and was associated with poor glycemic control (Karau et al. 2019). Haidari et al. (2016) reported similar prevalence of vitamin D deficiency among the Persian DM patients (35.72%). Vitamin D has protective effects through antagonizing the inflammatory response observed early during the pre-DM by reducing release of cytokines, chemokines, and reducing the monocytes chemotaxis (Bartels et al. 2010; Calton et al. 2015). It is also important for preserving the mitochondrial functions including maintaining the energetic one through preserving the function of the electron transport chain (ECT; Ashcroft et al. 2021). ECT is responsible for producing the majority of the energy needed by the cells; hence, mitochondria are called the powerhouse of the cells (Siekevitz 1957). Vitamin D antagonizes apoptosis (Moz et al. 2020; Riachy et al. 2002) and plays a pivotal role in controlling calcium hemostasis by increasing the expression of plasma calcium ATPase (Kip and Strehler 2004), NCX1, and calbindin (Ko et al. 2009; Pu et al. 2016) and decreasing expression of L-type calcium channels (Brewer et al. 2001). Together with its antioxidant effects (Wu et al. 2021), it regulates histone demethylase genes which control hypermethylation of promoter regions of many genes (Pereira et al. 2012; Yu et al. 2018).

Exposure to environmental contaminants, like pesticides, has hazardous effects. Organochlorine pesticides were introduced in the market and were extensively used around 80 years ago and were banned in the 1970s in many countries after many reports about their toxic effects (Blus 2002). A negative correlation existed between the serum concentration of organochlorine pesticides and vitamin D levels in a cross-sectional study that included 1275 participants (Yang et al. 2012). This toxic effect might be attributed to the endocrinal disruptor effects of these pesticides (Lee et al. 2010). Furthermore, the organochlorine pesticides interfere with vitamin D activity, its medicated intestinal absorption, intestinal alkaline phosphatase activity, and bone resorption (Nowicki et al. 1972).

Later, organophosphate pesticides were introduced in the market and as replacers of the banned organochloride ones (Costa 1987). In spite of the fact that many of scientific publications show their neurotoxic actions, they represent one of the most commonly used pesticides worldwide. Their neurotoxic actions could be attributed to inhibition of cholinesterase enzyme, mitochondrial dysfunction, oxidative stress, and inducing neuroinflammation (Jokanović 2018; Sakata 2005; Salama et al. 2014). To the extent of our knowledge, limited data exist regarding the effects of organophosphate pesticides on vitamin D. Chlorpyrifos is one of the most commonly used organophosphate pesticides. It increases the expression of VDR at the level of the skin which could indicate interference with vitamin D metabolism (Sawicki et al. 2019). Furthermore, vitamin D antagonized the chlorpyrifos-induced toxicity at retinal and renal levels of Wistar rats (El-Hossary et al. 2009).

12.4 Potential Association of Environmental Factors and PD, the Egyptian Experience

The occurrence of PD in some agricultural countries such as Egypt is higher than industrialized countries. Among 100,000 individuals, around 2748 individuals aged 50 years old or older and 7263 individuals aged between 70 and 79 years old are diagnosed with PD in Egypt (Khedr et al. 2012).

Only 5–15% of PD could be attributed to well-identified genetic mutation, while majority of PD patients are classified as idiopathic cases, postulated to be associated with exposure to environmental contaminants (e.g., pesticides, unhealthy lifestyle, and nutritional habits leading to micronutrients deficiency and central obesity) (Ball et al. 2019; Chen et al. 2004; Guo et al. 2019; Sherzai et al. 2016). These modifiable risk factors among the Egyptian population will be discussed in more details in the next section.

The health effects of exposure to pesticides among the Egyptian population have been fully reviewed by Mansour (2004 and 2008). While pesticides play a crucial role in increasing crop productivity, exposure to many pesticides can be associated with many health hazards, including metabolic and neurodegenerative disorders. Egypt was ranked as the fifth highest consumer of pesticides in Africa. Even though organochlorine pesticides like dichlorodiphenyltrichloroethane (DDT) have been banned in the 1970s, they are still detected in water and food. Furthermore, organophosphate pesticides, like chlorpyrifos and malathion, are one of the most widely used pesticides in Egypt. The problem of exposure to pesticides in Egypt is not only related to the amount of used pesticides but also to the lack of using protective equipment during mixing and application. We previously reported association between exposure to pesticides and risk of PD among the Egyptian population (OR: 7.09, 95% CI: 1.12–44.01). This risk is altered by polymorphism of the butyrylcholinesterase gene, which is responsible for metabolizing the organophosphate pesticides (Rösler et al. 2018).

DM increases the risk of developing PD, as was discussed in the previous section. Egypt is located in the red zone of DM, where more than half of diabetic patients reside (Al-Rubeaan 2010). The International Diabetes Federation reported that around 15.56% of Egyptian aged between 20 and 79 in 2011 were diabetic. In 2007, around 4.4 million Egyptian were diabetic, and the number increased to 7.5 and 8.9 million in 2013 and 2019, respectively. Furthermore, the number of diabetic patients in Egypt is expected to reach 11.9 and 16.9 million by 2030 and 2045, respectively (Aguiree et al. 2013; Karuranga et al. 2019; Whiting et al. 2011). We discussed the potential risk factors in our previous work, and this high prevalence of DM among the Egyptian population could be explained by unhealthy nutritional habits and lack of physical exercise which are one of the leading factors of the high prevalence of overweight and obesity among the Egyptian (Hegazi et al. 2015), as 50% and (65–80%) of Egyptian men and women, respectively, are overweight and obese (El-Zanaty and Way 2009). Furthermore, Sharara et al. (2018) show in their meta-analysis that the prevalence of physical inactivity among the Egyptian population is between 32% and 91%. Different types of exercise could offer neuroprotective effect and reduce PD progression at motor and non-motor levels. These protective effects could be attributed to the fact that exercise improves the mitochondrial function, antagonizes oxidative stress, enhances growth factor formation, positively affects neurogenesis and plasticity, and decreases the other modifiable risk factors like DM and cardiovascular diseases, as reviewed in detail by Feng et al. (2020) and Xu et al. (2019). Other potential risk factors for DM in Egypt are exposure to pesticides and the high prevalence of hepatitis C infection (Hegazi et al. 2015).

Regarding the status of vitamin D in Egypt, some of trails were conducted to assess the degrees of vitamin D insufficiency and deficiency among the Egyptian populations; however, small sample size was obvious among the majority of these studies. The prevalence of vitamin D insufficiency and deficiency among 200 Egyptian school students aged between 9 and 11 years old was 15 and 11.5%, respectively. Obesity and lack of adequate milk intake, sun exposure, and physical activity are linked to low vitamin D levels (Abu Shady et al. 2016). In one study, 24 and 21.3% of 75 adolescent Egyptian girls aged between 14 and 17 years old have insufficient and deficient vitamin D, respectively. This was associated with lack of sun exposure (Amr et al. 2012). Furthermore, El Badawy et al. (2015) showed that 18.5 and 5.3% of 500 Egyptian students between 13 and 18 years of age have insufficient and deficient vitamin D. Botros et al. (2015) reported in their cross-sectional study which included 404 Egyptian females that 77.2, 72.6, 72, and 39.5% of the geriatric, nursing, childbearing, and elderly females are suffering from vitamin D deficiency. Gerges et al. (2021) reported that 13 and 43% of 100 Egyptian females in the childbearing period have insufficient and deficient vitamin D levels. Another cross-sectional study assessed vitamin D levels among 135 pregnant Egyptian women and their neonates. Maternal levels of vitamin D were correlated to the levels in their neonates, as vitamin D insufficiency and deficiency were reported among 28.9 and 40% of the pregnant women and 32.6 and 60% of their neonates. Increased BMI and lack of fish intake and sun exposure were associated with low maternal vitamin D levels (El Rifai et al. 2014). As it was discussed in the previous section, vitamin D deficiency increases the risk of insulin resistance and in turn DM. The prevalence of vitamin D deficiency among 60 Egyptian patients with DM was reported by Abdelsadek et al. (2018) as 73.3% compared to 35% among 30 control individuals. Consistently, 450 obese Egyptian women aged between 25 and 35 years old were categorized according to the existence of vitamin D deficiency into two groups. Metabolic changes in the form of higher blood pressure, dyslipidemia, and insulin resistance with evidence of higher fasting blood glucose, and higher levels of inflammatory markers, C reactive protein, and interleukin 6, were obvious among vitamin D-deficient group. VDR gene polymorphisms alter the body response, by which women having ApaI (Aa, aa), FokI (Ff, ff), and for TaqI (Tt, tt) alleles have lower vitamin D levels, higher inflammatory markers, and more insulin resistance than those having common homozygous alleles ApaI (AA), FokI (FF), and TaqI (TT; Zaki et al. 2017).

12.5 Vitamin D as a Potential Therapy for PD

12.5.1 Experimental Studies

Experimental Parkinson's disease is modeled by different regimens of administration of toxins like 6-OHDA, MPTP, and rotenone, as reviewed recently by El-Gamal et al. (2021). Previous in vitro and in vivo studies reported protective effects of vitamin D against these toxins. Jang et al. (2014) show a neuroprotectant effect of vitamin D against rotenone-induced toxicity (reduced cell viability and oxidative stress) in SH-SY5Y cells, as in vitro model of PD, through activating the autophagy degradation pathway. Furthermore, Wang et al. (2001) reported the neuroprotective effect of vitamin D at both in vitro and in vivo levels. Daily administration of vitamin D to Sprague-Dawley rats for 8 days before stereotactic injection of 6-OHDA into the medial forebrain bundle improves the locomotor impairment and increase the dopamine and its metabolites in the substantia nigra. Coherently, vitamin D antagonized 6-OHDA-induced cell death of primary ventral mesencephalic culture. The protective effect of vitamin D in the 6-OHDA mouse model was further confirmed by Kim et al. (2020), as vitamin D administration antagonizes 6-OHDA dopaminergic neurodegeneration and neuroinflammation in substantia nigra. This beneficial effect could be explained that vitamin D reversed the 6-OHDA-induced reduction in the brain endothelial P-glycoprotein level and expression of VDR and its target genes MDR1a and CYP24. Similar results obtained among the 6-OHDA rat model of PD, as vitamin D, improves the locomotor impairment and dopamine content in the corpus striatum by antagonizing the inflammatory and oxidative stress processes (Lima et al. 2018). Together with that, vitamin D exerted a neuroprotective effect against MPTP-induced nigrostriatal neurodegeneration through reducing microglial activation, TLR4 receptor expression, pro-inflammatory cytokines expression, and increase expression of anti-inflammatory cytokines (IL-4, IL-10, and TGF-β) and CD (163, 204, and 206; Calvello et al. 2017).

Other potential mechanisms for the beneficial effects of vitamin D can be due to the fact that vitamin D plays a crucial role in proliferation and differentiation of neural stem cells (Shirazi et al. 2015) and increases the expression of tyrosine hydroxylase, the rate-limiting enzyme for DA synthesis (Cui et al. 2015). Furthermore, its deficiency is associated with impairment on the ontogeny of DA neurons in the developing brain (Cui et al. 2010). The potential molecular and environmental factors that could explain association of vitamin D and PD were illustrated in Fig. 12.1.

12.5.2 Clinical Studies

Furthermore, few clinical trials were conducted to assess the potential therapeutic role of vitamin D. Daily administration of vitamin D (10 000 IU) for 16 weeks in PD

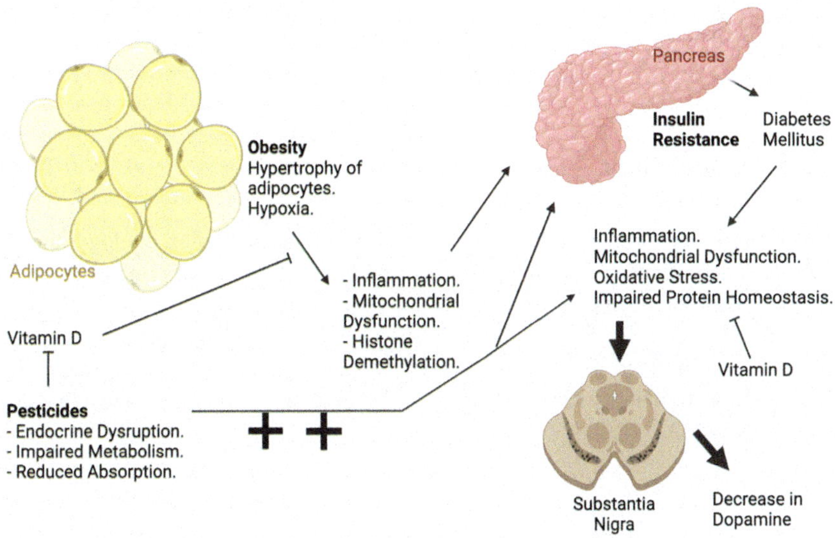

Fig. 12.1 The potential molecular and environmental factors that could explain association of vitamin D and PD

patients did not improve balance. However, a significant effect of vitamin D was reported in younger PD patients (aged 52 to 66 years old) but not the older PD patients (Hiller et al. 2018). Daily vitamin D supplementation (1200 IU) to PD patients for 1 year seems to be helpful in reducing the progression of the disease. However, this beneficial effect is variable according to different VDR Fokl genotype, as this effect was significant with Fokl CT and TT but not CC genotype (Suzuki et al. 2013). Zhou et al. (2019) demonstrated in their recent systematic review and meta-analysis that included eight studies that both vitamin D insufficiency and deficiency significantly increase the risk of PD (OR, 1.77; 95% CI, 1.29–2.43) and (OR, 2.55; 95% CI, 1.98–3.27), respectively. Exposure to sun for at least 15 minutes per week significantly reduces PD risk (OR, 0.02; 95% CI, 0.00–0.10). In spite of the fact that both vitamin D insufficiency and deficiency significantly increase the PD risk, vitamin D supplement did not significantly reduce motor impairment associated with PD.

With the declaration of the COVID-19 pandemic in the first half of 2020, it was suggested that COVID-19 can induce neurodegeneration and worsen PD manifestations. It has also been postulated that vitamin D administration could have a beneficial therapeutic effect (de Barros Viana et al. 2021). Hribar et al. (2020) concluded in their recent review that vitamin D supplements could have beneficial effects on reducing PD progression and both the risk and severity of COVID-19 among PD patients. Interestingly, a recent review of literature suggested that the association of low level of vitamin D among PD patients and severity of the PD motor symptoms could be attributed to limited mobility and lack of exposure to the sun as the disease progresses (Fullard and Duda 2020).

12.6 Conclusion

Like other noncommunicable chronic diseases, PD has complex etiology and risk factors. The current review suggests that vitamin D deficiency is one of these risk factors. This interesting association could be related to the pleiotropic effects of vitamin D especially its anti-inflammatory and antioxidant properties. Further research is warranted to better understand this association and may open the door for potential utilizing vitamin D as one of therapeutic modalities and preventive strategies of PD.

References

Abdelsadek SE, El Saghier EO, Abdel Raheem SI (2018) Serum 25(OH) vitamin D level and its relation to diabetic peripheral neuropathy in Egyptian patients with type 2 diabetes mellitus. Egypt J Neurol Psychiatry Neurosurg 54(1):36. https://doi.org/10.1186/s41983-018-0036-9

Abu Shady MM, Youssef MM, Salah El-Din EM, Abdel Samie OM, Megahed HS, Salem SME, Mohsen MA, Abdel Aziz A, El-Toukhy S (2016) Predictors of serum 25-Hydroxyvitamin D concentrations among a sample of Egyptian schoolchildren. Sci World J 2016. https://doi.org/10.1155/2016/8175768

Aguiree F, Brown A, Cho NH, Dahlquist G, Dodd S, Dunning T, Hirst M, Hwang C, Magliano D, Patterson C, Scott C, Shaw J, Soltesz G, Usher-Smith J, Whiting D (2013) IDF diabetes atlas, 6th edn

Al-Rubeaan K (2010) Type 2 diabetes mellitus red zone. Int J Diabetes Mellitus 2(1):1–2. https://doi.org/10.1016/j.ijdm.2009.12.009

Amr N, Hamid A, Sheta M, Elsedfy H (2012) Vitamin D status in healthy Egyptian adolescent girls. Georgian Med News 210:65–71. https://europepmc.org/article/med/23045423

Ashcroft SP, Fletcher G, Philp AM, Jenkinson C, Das S, Hansbro PM, Atherton PJ, Philp A (2021) Diet-induced vitamin D deficiency reduces skeletal muscle mitochondrial respiration. J Endocrinol 249(2):113–124. https://doi.org/10.1530/JOE-20-0233

Bach J-P, Ziegler U, Deuschl G, Dodel R, Doblhammer-Reiter G (2011) Projected numbers of people with movement disorders in the years 2030 and 2050. Mov Disord 26(12):2286–2290. https://doi.org/10.1002/mds.23878

Ball N, Teo W-P, Chandra S, Chapman J (2019) Parkinson's disease and the environment. Front Neurol 10:218. https://doi.org/10.3389/fneur.2019.00218

Bartels LE, Hvas CL, Agnholt J, Dahlerup JF, Agger R (2010) Human dendritic cell antigen presentation and chemotaxis are inhibited by intrinsic 25-hydroxy vitamin D activation. Int Immunopharmacol 10(8):922–928. https://doi.org/10.1016/j.intimp.2010.05.003

Berridge MJ (2017) Vitamin D deficiency and diabetes. Biochem J 474(8):1321–1332. https://doi.org/10.1042/BCJ20170042

Blus LJ (2002) Organochlorine pesticides. In: Handbook of ecotoxicology, 2nd edn. CRC Press, pp 313–339. https://doi.org/10.1201/9781351069656-1

Botros RM, Sabry IM, Abdelbaky RS, Eid YM, Nasr MS, Hendawy LM (2015) Vitamin D deficiency among healthy Egyptian females. Endocrinol Nutr 62(7):314–321. https://doi.org/10.1016/j.endonu.2015.03.010

Bovolenta TM, De Azevedo Silva SMC, Arb Saba R, Borges V, Ferraz HB, Felicio AC (2017) Systematic review and critical analysis of cost studies associated with Parkinson's disease. Parkinson's Dis 2017. https://doi.org/10.1155/2017/3410946

Brewer LD, Thibault V, Chen KC, Langub MC, Landfield PW, Porter NM (2001) Vitamin D hormone confers neuroprotection in parallel with downregulation of L-type calcium channel expression in hippocampal neurons. J Neurosci 21(1):98–108. https://doi.org/10.1523/jneurosci.21-01-00098.2001

Brockmann K, Apel A, Schulte C, Schneiderhan-Marra N, Pont-Sunyer C, Vilas D, Ruiz-Martinez-J, Langkamp M, Corvol JC, Cormier F, Knorpp T, Joos TO, Gasser T, Schüle B, Aasly JO, Foroud T, Marti-Masso JF, Brice A, Tolosa E et al (2016) Inflammatory profile in LRRK2-associated prodromal and clinical PD. J Neuroinflammation 13(1). https://doi.org/10.1186/s12974-016-0588-5

Butler MW, Burt A, Edwards TL, Zuchner S, Scott WK, Martin ER, Vance JM, Wang L (2011) Vitamin D receptor gene as a candidate gene for Parkinson disease. Ann Hum Genet 75(2): 201–210. https://doi.org/10.1111/j.1469-1809.2010.00631.x

Calton EK, Keane KN, Newsholme P, Soares MJ (2015) The impact of vitamin D levels on inflammatory status: a systematic review of immune cell studies. PLoS One 10(11):e0141770. https://doi.org/10.1371/journal.pone.0141770

Calvello R, Cianciulli A, Nicolardi G, De Nuccio F, Giannotti L, Salvatore R, Porro C, Trotta T, Panaro MA, Lofrumento DD (2017) Vitamin D treatment attenuates neuroinflammation and dopaminergic neurodegeneration in an animal model of Parkinson's disease, shifting M1 to M2 microglia responses. J Neuroimmune Pharmacol 12(2):327–339. https://doi.org/10.1007/s11481-016-9720-7

Chen H, Zhang SM, Schwarzschild MA, Hernán MA, Willett WC, Ascherio A (2004) Obesity and the risk of Parkinson's disease. Am J Epidemiol 159(6):547–555. https://doi.org/10.1093/aje/kwh059

Costa LG (1987) Toxicology of pesticides: a brief history. In: Costa LG, Galli CL, Murphy SD (eds) Toxicology of pesticides. Springer, Berlin Heidelberg, pp 1–10

Cui X, Pertile R, Liu P, Eyles DW (2015) Vitamin D regulates tyrosine hydroxylase expression: N-cadherin a possible mediator. Neuroscience 304:90–100. https://doi.org/10.1016/j.neuroscience.2015.07.048

Cui X, Pelekanos M, Burne THJ, McGrath JJ, Eyles DW (2010) Maternal vitamin D deficiency alters the expression of genes involved in dopamine specification in the developing rat mesencephalon. Neurosci Lett 486(3):220–223. https://doi.org/10.1016/j.neulet.2010.09.057

de Barros Viana M, Rosário BDA, de Fátima Santana de Nazaré M, Estadella D, Ribeiro DA, Socorro de Barros Viana G (2021) COVID-19 in age-related neurodegenerative diseases: is there a role for vitamin D3 as a possible therapeutic strategy? Rev Neurosci 32(2):235–247. https://doi.org/10.1515/revneuro-2020-0074

de Luca C, Olefsky JM (2008) Inflammation and insulin resistance. FEBS Lett 582(1):97–105. https://doi.org/10.1016/j.febslet.2007.11.057

Dowding CH, Shenton CL, Salek SS (2006) A review of the health-related quality of life and economic impact of Parkinson's disease. Drugs Aging 23(9):693–721. https://doi.org/10.2165/00002512-200623090-00001

Dursun E, Gezen-AK D, Yilmazer S (2014) Vitamin D Uygulamasının Primer Hippokampal Nöronlardaki İndüklenebilir Nitrik Oksit Sentaz (İNOS) Anlatımı Üzerine Etkileri. Nöro Psikiyatri Arşivi 51(2):163–168. https://doi.org/10.4274/npa.y7089

El-Gamal M, Salama M, Collins-Praino LE, Baetu I, Fathalla AM, Soliman AM, Mohamed W, Moustafa AA (2021) Neurotoxin-induced rodent models of Parkinson's disease: benefits and drawbacks. Neurotox Res:0123456789. https://doi.org/10.1007/s12640-021-00356-8

El-Hossary GG, El-Gohary AA, Ahmed NS, Mohamed AS, Mansour SM (2009) Amelioration of chlorpyrifos induced retinal and renal toxicity by vitamin D3. Aust J Basic Appl Sci 3(3): 2304–2314. https://www.researchgate.net/publication/288452421_Amelioration_of_Chlorpyrifos_Induced_Retinal_and_Renal_Toxicity_by_Vitamin_D

El-Zanaty F, Way A (2009) Egypt demographic and health survey 2008. https://www.unicef.org/egypt/reports/egypt-demographic-and-health-survey-2014

El Badawy AA, Aboserea MM, El Seifi OS, Mortada EM, Bakry HM, Waly EH, Raafat N, Etewa RL, El Badawy SA (2015) Vitamin D, parathormone and associated minerals among students in Zagazig district, Sharkia governorate, Egypt. Int J Vitam Nutr Res 84(3–4):173–182. https://doi.org/10.1024/0300-9831/a000204

El Rifai NM, Abdel Moety GAF, Gaafar HM, Hamed DA (2014) Vitamin D deficiency in Egyptian mothers and their neonates and possible related factors. J Matern Fetal Neonatal Med 27(10): 1064–1068. https://doi.org/10.3109/14767058.2013.849240

Feng YS, Yang SD, Tan ZX, Wang MM, Xing Y, Dong F, Zhang F (2020) The benefits and mechanisms of exercise training for Parkinson's disease. In: Life sciences, vol 245. Elsevier. https://doi.org/10.1016/j.lfs.2020.117345

Fullard ME, Duda JE (2020) A review of the relationship between vitamin D and Parkinson disease symptoms. In: Frontiers in neurology, vol 11. Frontiers Media S.A., p 454. https://doi.org/10.3389/fneur.2020.00454

Gao L, Tao Y, Zhang L, Jin Q (2010) Vitamin D receptor genetic polymorphisms and tuberculosis: updated systematic review and meta-analysis. Int J Tuberc Lung Dis 14(1):15–23

Gerges MES, Amin GEA, Andraous F, Abdel Hamid DM, Allam MF (2021) Vitamin D level in a sample of Egyptian females of childbearing age attending a family medicine center. Int J Clin Pract 75(4):e13738. https://doi.org/10.1111/ijcp.13738

Gozzelino R, Jeney V, Soares MP (2010) Mechanisms of cell protection by heme Oxygenase-1. In: Annual review of pharmacology and toxicology, vol 50, pp 323–354. https://doi.org/10.1146/annurev.pharmtox.010909.105600

Guo Y, Xu W, Liu FT, Li JQ, Cao XP, Tan L, Wang J, Yu JT (2019) Modifiable risk factors for cognitive impairment in Parkinson's disease: a systematic review and meta-analysis of prospective cohort studies. Mov Disord 34(6):876–883. https://doi.org/10.1002/mds.27665

Haidari F, Zakerkish M, Karandish M, Saki A, Pooraziz S (2016) Association between serum vitamin D level and glycemic and inflammatory markers in non-obese patients with type 2 diabetes. Iran J Med Sci 41(5):367–373

Han X, Xue L, Li Y, Chen B, Xie A (2012) Vitamin D receptor gene polymorphism and its association with Parkinson's disease in Chinese Han population. Neurosci Lett 525(1):29–33. https://doi.org/10.1016/j.neulet.2012.07.033

Hassan A, Sharma Kandel R, Mishra R, Gautam J, Alaref A, Jahan N (2020) Diabetes mellitus and Parkinson's disease: shared pathophysiological links and possible therapeutic implications. Cureus 12(8). https://doi.org/10.7759/cureus.9853

He J, Guo X, Liu ZQ, Yang PC, Yang S (2017) Vitamin D inhibits the staphylococcal enterotoxin B-induced expression of tumor necrosis factor in microglial cells. Immunol Res 65(4):913–919. https://doi.org/10.1007/s12026-017-8930-2

Hegazi RAF, Rao KN, Mayle A, Sepulveda AR, Otterbein LE, Plevy SE (2005) Carbon monoxide ameliorates chronic murine colitis through a heme oxygenase 1-dependent pathway. J Exp Med 202(12). https://doi.org/10.1084/jem.20051047

Hegazi R, El-Gamal M, Abdel-Hady N, Hamdy O (2015) Epidemiology of and risk factors for type 2 diabetes in Egypt. Ann Glob Health 81(6):814–820. https://doi.org/10.1016/j.aogh.2015.12.011

Heilbronn L, Campbell L (2008) Adipose tissue macrophages, low grade inflammation and insulin resistance in human obesity. Curr Pharm Des 14(12):1225–1230. https://doi.org/10.2174/138161208784246153

Hiller AL, Murchison CF, Lobb BM, O'Connor S, O'Connor M, Quinn JF (2018) A randomized, controlled pilot study of the effects of vitamin D supplementation on balance in Parkinson's disease: does age matter? PLoS One 13(9). https://doi.org/10.1371/journal.pone.0203637
Hirtz D, Thurman DJ, Gwinn-Hardy K, Mohamed M, Chaudhuri AR, Zalutsky R (2007) How common are the & quot; common& quot; neurologic disorders? Neurology 68(5):326–337. https://doi.org/10.1212/01.wnl.0000252807.38124.a3
Hribar CA, Cobbold PH, Church FC (2020) Potential role of vitamin d in the elderly to resist covid-19 and to slow progression of Parkinson's disease. Brain Sci 10(5):6–13. https://doi.org/10.3390/brainsci10050284
Jang W, Kim HJ, Li H, Jo KD, Lee MK, Song SH, Yang HO (2014) 1, 25-Dyhydroxyvitamin D3 attenuates rotenone-induced neurotoxicity in SH-SY5Y cells through induction of autophagy. Biochem Biophys Res Commun 451(1):142–147. https://doi.org/10.1016/j.bbrc.2014.07.081
Johnson SJ, Diener MD, Kaltenboeck A, Birnbaum HG, Siderowf AD (2013) An economic model of Parkinson's disease: implications for slowing progression in the United States. Mov Disord 28(3):319–326. https://doi.org/10.1002/mds.25328
Jokanović M (2018) Neurotoxic effects of organophosphorus pesticides and possible association with neurodegenerative diseases in man: a review. Toxicology 410(December):125–131. https://doi.org/10.1016/j.tox.2018.09.009
Karau PB, Kirna B, Amayo E, Joshi M, Ngare S, Muriira G (2019) The prevalence of vitamin D deficiency among patients with type 2 diabetes seen at a referral hospital in Kenya. Pan Afr Med J 34:1–11. https://doi.org/10.11604/pamj.2019.34.38.18936
Karuranga S, Malanda B, Saeedi P, Salpea P (2019) IDF atlas 9th edition and other resources. In: International diabetes federation, 9th edn https://www.diabetesatlas.org/en/resources/
Khedr EM, Al Attar GS, Kandil MR, Kamel NF, Abo Elfetoh N, Ahmed MA (2012) Epidemiological study and clinical profile of Parkinson's disease in the Assiut governorate, Egypt: a community-based study. Neuroepidemiology 38(3):154–163. https://doi.org/10.1159/000335701
Kim H, Shin JY, Lee YS, Yun SP, Maeng HJ, Lee Y (2020) Brain endothelial p-glycoprotein level is reduced in Parkinson's disease via a vitamin d receptor-dependent pathway. Int J Mol Sci 21(22):1–15. https://doi.org/10.3390/ijms21228538
Kip SN, Strehler EE (2004) Vitamin D3 upregulates plasma membrane Ca2+-ATPase expression and potentiates apico-basal Ca2+ flux in MDCK cells. Am J Physiol Renal Physiol 286(2): 363–369. https://doi.org/10.1152/ajprenal.00076.2003
Knekt P, Kilkkinen A, Rissanen H, Marniemi J, Sääksjärvi K, Heliövaara M (2010) Serum vitamin D and the risk of Parkinson disease. Arch Neurol 67(7):808–811. https://doi.org/10.1001/archneurol.2010.120
Ko S-H, Lee G-S, Vo TTB, Jung E-M, Choi K-C, Cheung K-W, Kim JW, Park J-G, Oh GT, Jeung E-B (2009) Dietary calcium and 1, 25-dihydroxyvitamin D3 regulate transcription of calcium transporter genes in Calbindin-D9k knockout mice. J Reprod Dev 55(2):137–142. https://doi.org/10.1262/jrd.20139
Koszowska AU, Nowak J, Dittfeld A, Brończyk-Puzoń A, Kulpok A, Zubelewicz-Szkodzińska B (2014) Obesity, adipose tissue function and the role of vitamin D. Central Eur J Immunol 39(2): 260–264. https://doi.org/10.5114/ceji.2014.43732
Kowal SL, Dall TM, Chakrabarti R, Storm MV, Jain A (2013) The current and projected economic burden of Parkinson's disease in the United States. Mov Disord 28(3):311–318. https://doi.org/10.1002/mds.25292
Kuno R, Wang J, Kawanokuchi J, Takeuchi H, Mizuno T, Suzumura A (2005) Autocrine activation of microglia by tumor necrosis factor-α. J Neuroimmunol 162(1–2):89–96. https://doi.org/10.1016/j.jneuroim.2005.01.015
Lee DH, Steffes MW, Sjödin A, Jones RS, Needham LL, Jacobs DR (2010) Low dose of some persistent organic pollutants predicts type 2 diabetes: a nested case-control study. Environ Health Perspect 118(9):1235–1242. https://doi.org/10.1289/ehp.0901480

Lima LAR, Lopes MJP, Costa RO, Lima FAV, Neves KRT, Calou IBF, Andrade GM, Viana GSB (2018) Vitamin D protects dopaminergic neurons against neuroinflammation and oxidative stress in hemiparkinsonian rats. J Neuroinflammation 15(1):1–11. https://doi.org/10.1186/s12974-018-1266-6

Lin CH, Chen KH, Chen ML, Lin HI, Wu RM (2014) Vitamin D receptor genetic variants and Parkinson's disease in a Taiwanese population. Neurobiol Aging 35(5):1212.e11–1212.e13. https://doi.org/10.1016/j.neurobiolaging.2013.10.094

Luo X, Ou R, Dutta R, Tian Y, Xiong H, Shang H (2018) Association between serum Vitamin D levels and Parkinson's disease: a systematic review and meta-analysis. Front Neurol 9 (Nov):1–10. https://doi.org/10.3389/fneur.2018.00909

Luo X, Lee Kraus W (2012) On par with PARP: cellular stress signaling through poly(ADP-ribose) and PARP-1. Genes Dev 26(5):417–432. https://doi.org/10.1101/gad.183509.111

Lương K, Nguyễn L (2012) Role of vitamin D in Parkinson's disease. ISRN Neurol 2012:1–11. https://doi.org/10.5402/2012/134289

Lv Z, Qi H, Wang L, Fan X, Han F, Wang H, Bi S (2014) Vitamin D status and Parkinson's disease: a systematic review and meta-analysis. Neurol Sci 35(11):1723–1730. https://doi.org/10.1007/s10072-014-1821-6

Mansour SA (2008) Environmental impact of pesticides in Egypt. Rev Environ Contam Toxicol 196:1–51. https://doi.org/10.1007/978-0-387-78444-1_1

Mansour S a (2004) Pesticide exposure--Egyptian scene. Toxicology 198(1–3):91–115. https://doi.org/10.1016/j.tox.2004.01.036

Martinez-Martin P, Macaulay D, Jalundhwala YJ, Mu F, Ohashi E, Marshall T, Sail K (2019) The long-term direct and indirect economic burden among Parkinson's disease caregivers in the United States. Mov Disord 34(2):236–245. https://doi.org/10.1002/mds.27579

Moz S, Contran N, Facco M, Trimarco V, Plebani M, Basso D (2020) Vitamin D prevents pancreatic cancer-induced apoptosis signaling of inflammatory cells. Biomol Ther 10(7):1–16. https://doi.org/10.3390/biom10071055

Nagatsu T, Sawada M (2005) Inflammatory process in Parkinsons disease: role for cytokines. Curr Pharm Des 11(8):999–1016. https://doi.org/10.2174/1381612053381620

Nowicki HG, Wong RG, Myrtle JF, Norman AW (1972) Inhibition of biological activity of cholecalciferol (vitamin D3) by o, p'-DDT or, p, p'-DDT in rachitic cockerel. J Agric Food Chem 20(2):376–380. https://doi.org/10.1021/jf60180a058

Oertel WH (2017) Recent advances in treating Parkinson's disease. F1000Research 6:260. https://doi.org/10.12688/f1000research.10100.1

Pereira F, Barbáchano A, Singh PK, Campbell MJ, Muñoz A, Larriba MJ (2012) Vitamin D has wide regulatory effects on histone demethylase genes. Cell Cycle 11(6):1081–1089. https://doi.org/10.4161/cc.11.6.19508

Peritore CS, Ho A, Yamamoto BK, Schaus SE (2012) Resveratrol attenuates L-DOPA-induced hydrogen peroxide toxicity in neuronal cells. Neuro Report 23(17):989–994. https://doi.org/10.1097/WNR.0b013e32835a4ea4

Phillips JR, Eissa AM, Hewedi DH, Jahanshahi M, El-Gamal M, Keri S, Moustafa AA (2016) Neural substrates and potential treatments for levodopa-induced dyskinesias in Parkinson's disease. Rev Neurosci. https://doi.org/10.1515/revneuro-2016-0009

Pu F, Chen N, Xue S (2016) Calcium intake, calcium homeostasis and health. Food Sci Human Wellness 5(1):8–16. https://doi.org/10.1016/j.fshw.2016.01.001

Reusch JEB (2002) Current concepts in insulin resistance, type 2 diabetes mellitus, and the metabolic syndrome. Am J Cardiol 90(5 SUPPL):19–26. https://doi.org/10.1016/S0002-9149(02)02555-9

Riachy R, Vandewalle B, Conte JK, Moerman E, Sacchetti P, Lukowiak B, Gmyr V, Bouckenooghe T, Dubois M, Pattou F (2002) 1, 25-dihydroxyvitamin D3 protects RINm5F and human islet cells against cytokine-induced apoptosis: implication of the antiapoptotic protein A20. Endocrinology 143(12):4809–4819. https://doi.org/10.1210/en.2002-220449

Rimmelzwaan LM, Van Schoor NM, Lips P, Berendse HW, Eekhoff EMW (2016) Systematic review of the relationship between vitamin D and Parkinson's disease. J Parkinsons Dis 6(1): 29–37). IOS press. https://doi.org/10.3233/JPD-150615

Rösler TW, Salama M, Shalash AS, Khedr EM, El-Tantawy A, Fawi G, El-Motayam A, El-Seidy E, El-Sherif M, El-Gamal M, Moharram M, El-Kattan M, Abdel-Naby M, Ashour S, Müller U, Dempfle A, Kuhlenbäumer G, Höglinger GU (2018) K-variant BCHE and pesticide exposure: gene-environment interactions in a case–control study of Parkinson's disease in Egypt. Sci Rep 8(1):16525. https://doi.org/10.1038/s41598-018-35003-4

Rossi A, Berger K, Chen H, Leslie D, Mailman RB, Huang X (2018) Projection of the prevalence of Parkinson's disease in the coming decades: revisited. Mov Disord 33(1):156–159. https://doi.org/10.1002/mds.27063

Sakata M (2005) II.7.2 Organophosphorus pesticides, pp 534–544. http://eknygos.lsmuni.lt/springer/124/535-544.pdf

Salama M, El-Morsy D, El-Gamal M, Shabka O, Mohamed WMY (2014) Mitochondrial complex I inhibition as a possible mechanism of chlorpyrifos induced neurotoxicity. Ann Neurosci 21(3): 85–89. https://doi.org/10.5214/ans.0972.7531.210303

Sawicki K, Czajka M, Matysiak-Kucharek M, Kruszewski M, Skawiński W, Brzóska K, Kapka-Skrzypczak L (2019) Chlorpyrifos stimulates expression of vitamin D 3 receptor in skin cells irradiated with UVB. Pestic Biochem Physiol 154:17–22. https://doi.org/10.1016/j.pestbp.2018.12.003

Sharara E, Akik C, Ghattas H, Makhlouf Obermeyer C (2018) Physical inactivity, gender and culture in Arab countries: a systematic assessment of the literature. BMC Public Health 18(1): 1–19. https://doi.org/10.1186/s12889-018-5472-z

Shen L, Ji HF (2015) Associations between vitamin D status, supplementation, outdoor work and risk of Parkinson's disease: a meta-analysis assessment. Nutrients 7(6):4817–4827. https://doi.org/10.3390/nu7064817

Sherzai AZ, Tagliati M, Park K, Pezeshkian S, Sherzai D (2016) Micronutrients and risk of Parkinson's disease. Gerontol Geriatr Med 2. https://doi.org/10.1177/2333721416644286

Shirazi HA, Rasouli J, Ciric B, Rostami A, Zhang G-X (2015) 1,25-Dihydroxyvitamin D3 enhances neural stem cell proliferation and oligodendrocyte differentiation. Exp Mol Pathol 98(2): 240–245. https://doi.org/10.1016/j.yexmp.2015.02.004

Shrestha S, Lutsey PL, Alonso A, Huang X, Mosley TH, Chen H (2016) Serum 25-hydroxyvitamin D concentrations in mid-adulthood and Parkinson's disease risk. Mov Disord 31(7):972–978. https://doi.org/10.1002/mds.26573

Siekevitz P (1957) Powerhouse of the cell. Sci Am 197(1):131–144. https://doi.org/10.1038/scientificamerican0757-131

Smith MP, Fletcher-Turner A, Yurek DM, Cass WA (2006) Calcitriol protection against dopamine loss induced by intracerebroventricular administration of 6-hydroxydopamine. Neurochem Res 31(4):533–539. https://doi.org/10.1007/s11064-006-9048-4

Stevens SL (2021) Fat-soluble vitamins. Nurs Clin N Am 56(1):33–45. https://doi.org/10.1016/j.cnur.2020.10.003

Suzuki M, Yoshioka M, Hashimoto M, Murakami M, Noya M, Takahashi D (2013) Randomized, double-blind, placebo-controlled trial of vitamin D. Am J Clin Nutr 97(3):1004–1013. https://doi.org/10.1007/s11481-016-9720-7

Szymczak-Pajor I, Śliwińska A (2019) Analysis of association between vitamin d deficiency and insulin resistance. Nutrients 11(4):794. https://doi.org/10.3390/nu11040794

Taylor SI, Accili D, Imai Y (1994) Insulin resistance or insulin deficiency: which is the primary cause of NIDDM? Diabetes 43(6):735–740. https://doi.org/10.2337/diab.43.6.735

Török R, Török N, Szalardy L, Plangar I, Szolnoki Z, Somogyvari F, Vecsei L, Klivenyi P (2013) Association of vitamin D receptor gene polymorphisms and Parkinson's disease in Hungarians. Neurosci Lett 551:70–74. https://doi.org/10.1016/j.neulet.2013.07.014

Toulouse A, Sullivan AM (2008) Progress in Parkinson's disease—where do we stand? Prog Neurobiol 85(4):376–392. https://doi.org/10.1016/J.PNEUROBIO.2008.05.003

Wang JY, Wu JN, Cherng TL, Hoffer BJ, Chen HH, Borlongan CV, Wang Y (2001) Vitamin D3 attenuates 6-hydroxydopamine-induced neurotoxicity in rats. Brain Res 904(1):67–75. https://doi.org/10.1016/S0006-8993(01)02450-7

Whiting DR, Guariguata L, Weil C, Shaw J (2011) IDF diabetes atlas: global estimates of the prevalence of diabetes for 2011 and 2030. Diabetes Res Clin Pract 94(3):311–321. https://doi.org/10.1016/j.diabres.2011.10.029

Wu M, Wu Y, Xu K, Lin L (2021) Protective effects of 1,25 dihydroxyvitamin D3 against high-glucose-induced damage in human umbilical vein endothelial cells involve activation of Nrf2 antioxidant signaling. J Vasc Res:1–10. https://doi.org/10.1159/000515512

Xu X, Fu Z, Le W (2019) Exercise and Parkinson's disease. Int Rev Neurobiol 147:45–74. https://doi.org/10.1016/bs.irn.2019.06.003

Yan J, Fu Q, Cheng L, Zhai M, Wu W, Huang L, Du G (2014) Inflammatory response in Parkinson's disease (review). Mol Med Rep 10(5):2223–2233. https://doi.org/10.3892/mmr.2014.2563

Yang JH, Lee YM, Bae SG, Jacobs DR, Lee DH (2012) Associations between organochlorine pesticides and vitamin d deficiency in the U.S. population. PLoS One 7(1):e30093. https://doi.org/10.1371/journal.pone.0030093

Yao L, Dai X, Sun Y, Wang Y, Yang Q, Chen X, Liu Y, Zhang L, Xie W, Liu J (2018) Inhibition of transcription factor SP1 produces neuroprotective effects through decreasing MAO B activity in MPTP/MPP+ Parkinson's disease models. J Neurosci Res 96(10):1663–1676. https://doi.org/10.1002/jnr.24266

Yu S, Wang Y, Li X, Mao Z, Yu F, Wang L, Ba Y, Wang C, Li W (2018) Methylation in 3′ near region of GC gene and its association with the level of vitamin D binding protein and type 2 diabetes mellitus. Nutr Res 54:52–59. https://doi.org/10.1016/j.nutres.2018.03.016

Zaki M, Kamal S, Basha WA, Youness E, Ezzat W, El-Bassyouni H, Amr K (2017) Association of vitamin D receptor gene polymorphism (VDR) with vitamin D deficiency, metabolic and inflammatory markers in Egyptian obese women. Genes Dis 4(3):176–182. https://doi.org/10.1016/j.gendis.2017.07.002

Zhou Z, Zhou R, Zhang Z, Li K (2019) The association between vitamin D status, vitamin D supplementation, sunlight exposure, and parkinson's disease: a systematic review and meta-analysis. Med Sci Monit 25:666–674. https://doi.org/10.12659/MSM.912840

Chapter 13
Crosstalk Between Autophagy and Nutrigenomics in Neurodegenerative Diseases

Ashley E. Russell, Asmaa Badr, Ali Rai, Cierra Carafice, Ansaar Rai, Kylene P. Daily, Owen Whitham, Amal O. Amer, Candice M. Brown, and Duaa Dakhlallah

13.1 Introduction

Proteostasis refers to the dynamic regulation of protein homeostasis which is mediated by a network of molecular machines responsible for both protein synthesis and degradation. With age and various diseases, proteostasis can be disrupted, leading to the formation of intracellular protein aggregates (Labbadia and Morimoto 2015).

Amal O. Amer and Candice M. Brown contributed equally with all other contributors.

A. E. Russell
Department of Biology, School of Science, Penn State Erie, The Behrend College, Erie, PA, USA

A. Badr · C. Carafice · K. P. Daily · O. Whitham · A. O. Amer
Department of Microbial Infection and Immunity, Infectious Diseases Institute, The Ohio State University, Columbus, OH, USA

A. Rai
Benjamin M Statler, College of Engineering, Chemical and Biomedical Engineering Department, West Virginia University, Morgantown, WV, USA

A. Rai
Department of Neuroradiology, Health Science Center, West Virginia University, Morgantown, WV, USA

C. M. Brown
Department of Neuroscience, Center for Basic and Translational Stroke Research, Rockefeller Neuroscience Institute, West Virginia University School of Medicine, Morgantown, WV, USA

D. Dakhlallah (✉)
Institute of Global Health and Human Ecology, School of Sciences & Engineering, The American University of Cairo, New Cairo, Egypt

Microbiology, Immunology and Cell Biology, and Mary Babb Randolph Cancer Center, Erma Byrd Biomedical Research Center, West Virginia University, Morgantown, WV, USA
e-mail: DUAA.DAKHLALLAH@aucegypt.edu

M. Salama (ed.), *Nutrigenomics and the Brain*, Nutritional Neurosciences,
https://doi.org/10.1007/978-981-16-9205-5_13

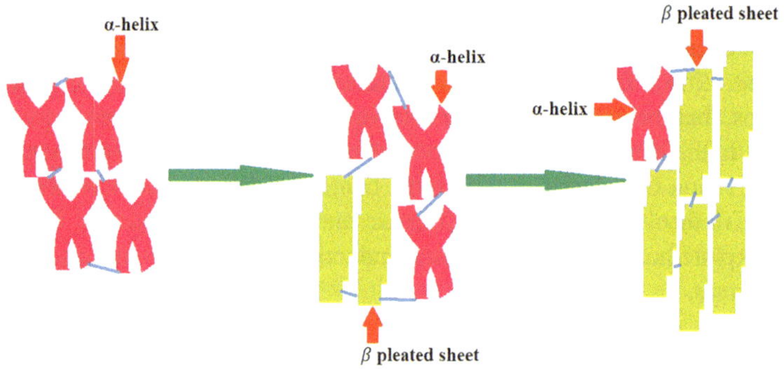

Fig. 13.1 The process by which an α-helix starts to form into a β-sheet, until the protein is mostly formed out of β-sheets

Tau is a microtubule-associated protein abundant in neuronal axons which controls microtubule stability based on its phosphorylation state. In AD, tau becomes hyperphosphorylated and does not bind and stabilize microtubules; instead it forms tau oligomers and neurofibrillary tangles. Microtubule instability impairs axonal transport leading to neuronal demise. Microglia and astrocytes are chronically activated in response to these misfolded proteins and instate a hostile inflammatory environment (Di Meco et al. 2020).

Apolipoprotein E (ApoE) is the largest genetic risk factor for AD. There are three ApoE protein isoforms in humans, i.e., ApoE2, ApoE3, and ApoE4; each one is encoded by one of three APOE alleles. While *APOE3* is the most common allele in the human population, inheritance of the *APOE4* allele dose-dependently increases the age of onset and severity of AD; in contrast, inheritance of the *APOE2* allele may confer neuroprotection (Long and Holtzman 2019; Pericak-Vance and Haines 1995; Strittmatter and Roses 1996). There are many multiple presumed mechanisms through which the ApoE4 is thought to contribute to the AD risk such as an increase in Aβ deposition, formation of neurofibrillary tangles, and neuroinflammation. There are also numerous other genetic risk factors for AD identified through genome-wide association studies such as variants of TREM2, SORL1, and ABCA7 (Scheltens et al. 2021).

Numerous studies have demonstrated that there is precipitous decline in autophagy during AD. Disrupted autophagy in AD is largely characterized by the presence of immature autophagic vacuoles that accumulate inside diseased neurons (Nixon et al. 2005). In an animal model of AD, a mutation in the presenilin 1 gene (PSEN1) associated with AD onset in humans has been linked to autophagy deficits (Chong et al. 2018). Dysfunctional autophagy is also implicated in failed neuronal Aβ metabolism and degradation (Li et al. 2017; Eshraghi et al. 2021). Furthermore, Aβ is thought to be generated in neuronal lysosomes, endoplasmic reticulum, or Golgi apparatus, but there is evidence that autophagy may also contribute to the generation of Aβ (Mizushima 2005). The accumulation of autophagic vacuoles in AD neurons could serve as a source of continued Aβ generation (Nixon et al. 2005).

In microglia, autophagy participates in Aβ degradation, and weak autophagy activity is associated with failed Aβ clearance and unsuccessful regulation of neuro-inflammatory phenotypes (Lucin et al. 2013; Heckmann et al. 2019; Houtman et al. 2019; Shibuya et al. 2014; Cho et al. 2014; Liu et al. 2020). Furthermore, a role for noncanonical autophagy has been described for microglial phagocytosis of Aβ (Heckmann et al. 2019). Stimulation of autophagy is also associated with a reduction in tau levels (Congdon et al. 2012). Lastly, recent data suggests that ApoE4 is also linked to autophagy dysfunction. APOE4 carriers have also been shown to exhibit decreased autophagy through a direct interaction with autophagy proteins or a potential transcription factor (Kloske and Wilcock 2020). The interaction between other genetic risk factors and autophagy is less clear and is an important area for further study.

13.2.1 *Metabolic Conditions Associated with Alzheimer's Disease*

Many studies link specific metabolic diseases to a higher incidence of AD. Notably, these metabolic diseases are associated with dysfunctional autophagy. Three of these metabolic disorders are discussed below.

13.2.1.1 Diabetes Mellitus

Cross talk in the pathophysiological processes causing neurodegenerative disorders and metabolic diseases have been extensively investigated, in which dysfunctional autophagy and insulin signaling disturbance appear to be a common factor in both conditions. Type 2 diabetes mellitus (DM) is associated with increased incidence of neurodegenerative diseases such as AD and dementia and the decline in cognitive ability (De Mello et al. 2019; Burillo et al. 2021; MacKnight et al. 2002; Arvanitakis et al. 2004). Insulin is being secreted into the cerebrospinal fluid (CSF); albeit its concentration is only 27% that of the insulin concentration in blood plasma (Akomolafe et al. 2006; Begg 2015; Margolis and Altszuler 1967). Additionally, insulin was found to cross the blood-brain barrier through a specific transport system (Begg 2015; Schwartz et al. 1991). Choroid plexus and astrocytes are a putative source of insulin (Mazucanti et al. 2019).

The role of insulin and insulin receptors in promoting neuronal plasticity and cognitive functions is well established (Akomolafe et al. 2006; Grote and Wright 2016). Neuronal growth and axonal guidance are enhanced by insulin through the activation of PI3K/AKT pathway, which also controls the autophagy process. Both Aβ and lipopolysaccharide (LPS) cause significant decreases in the level of insulin secretion by cultured astrocytes (Takano et al. 2018). A constitutively active autophagy mouse model, produced by induction of Beclin1 point mutation, showed increased insulin sensitivity with the consumption of high fat diet, despite the increase in insulin granules degradation (De Mello et al. 2019; Yamamoto et al. 2018). Furthermore, crossing this constitutively active autophagy mouse model Becn1F121A with the Alzheimer's disease mouse model 5XFAD lead to lower levels of both soluble and insoluble Aβ in the cortex and hippocampus and prevented cognitive decline (Rocchi et al. 2017).

Interestingly, the use of the antidiabetic drug metformin has shown positive effects on neurodegenerative diseases in animal models (Katila et al. 2017; Lu et al. 2016). Both acute and chronic metformin administration caused increases in glucagon-like peptide-1 (GLP-1) levels with subsequent enhancement in insulin secretion. This effect caused activation of PI3K/AKT signaling, accompanied by higher ATG7 in the brain, and autophagy stimulation (Lu et al. 2016; Patil et al. 2014; Ghadernezhad et al. 2016). Additionally, insulin and insulin-like growth factor 1 (IGF-1) bind and activate insulin receptor substrate 1 (IRS-1), leading to activation of PI3K and generation of PI3P with subsequent autophagy stimulation (Candeias et al. 2017; Lapchak and Araujo 2001). Therefore, AD is now sometimes referred to as "Type 3 diabetes," reflecting a new mechanism of neurodegeneration (Steen et al. 2005).

13.2.1.2 Hypercholesterolemia

High blood cholesterol levels have been linked to an increase in the risk of developing AD later in life (Peric and Annaert 2015; Huang and Mahley 2014; Lahiri et al. 2004). One of the main genetic risk factors for AD is the apolipoprotein E4 (APOE4) allele (Liao et al. 2017). ApoE helps maintain neuronal structures and enhances synaptic repair (Kim et al. 2014). Nevertheless, people that carry the ApoE4 mutation are more susceptible to the development of AD (Lahiri et al. 2004; Beyer et al. 2005). This mutation is known to increase the circulation levels of low density lipoprotein (LDL). High LDL-C levels are associated with increase in the risk for developing AD. Importantly, APOE4 enhances Aβ accumulation by binding to 12–28 residues of Aβ and worsens oxidative stress associated with AD (Huang and Mahley 2014; Liao et al. 2017; Kim et al. 2009; Potter and Wisniewski 2012). Neuronal death in the hippocampus in APOE4 transgenic mice was associated with elevated lysosomal Aβ42 (Belinson et al. 2008). Additionally, APOE4 causes lysosomal leakage in neuro-2a cells (Ji et al. 2006), and APOE4 is also known to cause autophagy dysfunction (Van Acker et al. 2019; Simonovitch et al. 2016). Interestingly, statin, a cholesterol level-lowering medication, reduces levels of Aβ and decreases in the risk of developing AD both in mice and most importantly humans (Arvanitakis et al. 2008; Shinohara et al. 2014; Li et al. 2018).

13.2.1.3 Oxidative Stress

The neurons with the highest metabolic rate in the brain are particularly susceptible to oxidative damage by reactive oxygen and nitrogen species. AD neurons contain a large amount of unsaturated fatty acids that undergo peroxidation and subsequent destruction by reactive oxygen species (ROS) (Pocernich and Butterfield 2012). Additionally, they express low levels of the antioxidant enzyme glutathione, which clears free radicals. Thus, neurons are vulnerable to the effect of oxidative stress, which is a hallmark of AD (Butterfield and Kanski 2001; Beal 2002); there are multiple serological markers for oxidative stress that are associated with AD such as 8-hydroxydeoxyguanosine (8-OHdG) (Zhao and Zhao 2013), 8-hydroxyguanosine, 4-hydroxynonenal (8-OHG), and malondialdehyde (MDA) (Lovell et al. 1995; Williams et al. 2006). Moreover, the suppression of expression and activity of some antioxidant enzymes, such as superoxide dismutase (SOD) and catalase, have also been described in AD patients (Marcus et al. 1998). The mechanisms underlying the production of ROS and oxidative stress include the accumulation of hyperphosphorylated tau and Aβ, in addition to mitochondrial dysfunction (Cha et al. 2015; Chen and Zhong 2014).

Oxidative stress largely contributes to accumulation of Aβ (Shinohara et al. 2014; Chen and Zhong 2014; Butterfield et al. 2013) and tau hyperphosphorylation (Alavi Naini and Soussi-Yanicostas 2015). Moreover, oxidative stress contributes to the suppression of autophagy which accompanies AD (Liu et al. 2015). Additionally, the activation of autophagy offered protection against neurotoxicity caused by oxidative stress caused by thiamin deficiency (Karuppagounder et al. 2009). This vicious circle of oxidative stress-autophagy deficiency and AD can be interrupted by prevention of oxidative stress using antioxidants. Finally, various antioxidants including vitamin E and C can protect against neurotoxicity and serve as biomarkers and treatment targets for AD (Galasko et al. 2012).

13.2.2 Nutrition-Induced Autophagy Activation as a Potential AD Therapy

Inducing autophagy has been a major target of interest in the treatment of AD. Most notably, rapamycin, which inhibits mTOR, has entered clinical trials as it decreased Aβ and tau pathology in preclinical research (Caccamo et al. 2014; Spilman et al. 2010; Zhang et al. 2017; Yang and Zhang 2020). Several compounds found in dietary substances have also been shown to improve autophagy. Many of these compounds have been studied specifically in the context of AD. The general mechanism of autophagy improvement of notable examples studied for use in AD is listed in Table 13.1. References included offer mechanisms underlying autophagy improvement and select studies completed in AD models. Many of the foods which

Table 13.1 Food-based autophagy-improving compounds shown to potentially modulate AD pathobiology (CRM: Caloric Restriction Mimetic)

Compound	Dietary source	Mechanisms of autophagy improvement in AD models
Oleuropein aglycone (Fernández-Sanz et al. 2019; Cordero et al. 2018)	Extra virgin olive oil (Mediterranean diet)	• ↑expression of Beclin-1, cathepsin B, p62, and LC3 • Inhibition of mTORC • Activation of sirtuins like SIRT1
Quercetin (Fernández-Sanz et al. 2019; Sabogal-Guáqueta et al. 2015; Costa et al. 2016)	Capers, dark chocolate, red onions, cloves	• Activation of SIRT1
Trehalose (Fernández-Sanz et al. 2019; Du et al. 2013)	Mushrooms	• ↑formation of autophagosomes, independent of mTOR inhibition
Resveratrol (Deng and Mi 2016; Sun et al. 2019; Vingtdeux et al. 2010)	Peanuts, pistachios, grapes, wine, blueberries, cranberries	• CRM • Activation of SIRT1 • Repression of mTOR signaling by activating AMPK signaling
Spermidine (Pietrocola et al. 2015; Wirth et al. 2018; Sandusky-Beltran et al. 2019)	Green pepper, wheat germ, cauliflower, broccoli, mushrooms, cheeses, soybeans	• CRM • Inhibits acetyltransferase EP300 which deacetylates cellular autophagy related proteins • ↑expression of autophagy-related genes by altering epigenetic mechanisms
Aspirin (Pietrocola et al. 2018; Chandra et al. 2018)	Several fruits, vegetables, and herbs	• CRM • Competing with acetyl coenzyme A to block EP300 • ↑lysosomal biogenesis via activation of transcription factor EB (TFEB)
Curcumin (Reddy et al. 2018; Shakeri et al. 2019; Wang et al. 2014)	Found in turmeric	• CRM • Inhibiting mTOR signaling

contain these compounds are part of a normal healthy diet or are found in well-established beneficial dietary patterns, such as the Mediterranean diet.

Caloric restriction is suggested to improve autophagy function in aging, neurodegeneration, and AD (Ntsapi et al. 2019). The most common form of caloric restriction is intermittent fasting, which involves alternating periods of access to food and restriction. This dietary practice is thought to induce autophagy via induction of AMPK and sirtuin-1 (SIRT1) or inhibition of the insulin/IGF1 pathway and mTORC1 signaling (Pani 2015). Several studies have shown potential benefit of caloric restriction in aging and AD mice (Schafer et al. 2015; Bondolfi et al. 2004; Patel et al. 2005). However, more work is needed to understand the optimal caloric restriction regimen in humans, including long-term studies. Several challenges will be important to consider in such studies, such as sex-based differences, in response to caloric restriction and compliance issues with this dietary regimen (Ntsapi et al. 2019; Martin et al. 2007). In light of these potential challenges, there are also dietary compounds considered caloric restriction mimetics (CRMs) that may be beneficial (Yang and Zhang 2020). These CRMs have been studied in the context of AD and are also listed in Table 13.1.

Further research is needed to understand what dietary modifications are beneficial in AD patients. Additionally, dosage requirements and potential toxicity of these compounds need to be explored. Notably, some medical foods are already approved for treatment of AD patients. These foods are comprised of combinations of fatty acids and vitamins (Masters et al. 2015). Certainly, the potential of dietary modifications has gained traction in the treatment of AD patients. The benefits of diet in improving autophagy in AD warrant further investigation.

There is clear potential for modulating autophagy in AD by altering diet and nutrition. Autophagy is weak in AD, and bolstering its function is an exciting target for clinical trials. Several dietary compounds known to stimulate autophagy have shown promise in preclinical research. Additionally, a more robust understanding of the links between certain metabolic diseases and AD may allow for targeted prevention efforts. Though targeting autophagy in AD with diet is promising, clinical trials are necessary to identify clear regimens, safety profiles, and characteristics of patients responding to treatment.

13.3 Parkinson's Disease

Parkinson's disease (PD) is the second most common neurodegenerative disease. The incidence of PD increases with age, and it is therefore most common in individuals older than 65. The most prominent physical symptoms of PD are rest tremors, bradykinesia, postural and gait impairment, and muscular rigidity, while nonphysical symptoms include cognitive decline, depression, and sleeping disorders. The hallmark neuropathology of PD is characterized by the loss of dopaminergic neurons in the substantia nigra pars compacta region of the brain, which accompanied by the accumulation of Lewy bodies in dopaminergic neurons. Several mechanisms have been implicated in dopaminergic (DA) neuron degeneration, including mitochondrial dysfunction, oxidative stress, and autophagy (Dionísio et al. 2021; Kalia and Lang 2015). Studying the genetics of PD has led to important mechanistic insights on the role of autophagy in disease pathophysiology. In this section, we will review this evidence and demonstrate that nutritional intervention may be a way to limit autophagy and prevent disease progression.

PD has a known genetic component to the disease, as 5–10% of patients have the familial form of parkinsonism with autosomal-dominant pattern of inheritance. However, patients can also have idiopathic sporadic PD, in which environmental factors are emphasized as causative agents (Esposito et al. 2002) . An extensive range of endogenous and exogenous factors play a role in PD, including specific gene mutations and environmental factors like diet. Other genes linked to mitochondrial dysregulation that cause PD include *LRRK2*, *DH1*, *ATP13A2*, and *SCNA*. Additionally, one of the major causes of early-onset PD is due to loss-of-function mutations in glucocerebrosidase (*GBA*), *RAB39B*, *DJ-1*, and *PARKIN* (Quinn et al. 2020). Epidemiological evidence has shown some plausible risk factors and potential protective factors for PD. Risk factors include pesticides, dairy products, and B2-adrenoreceptor antagonists, while potential protective factors are caffeine and tea intake, physical activity, and vitamin E intake. The protective factors have been shown to slow down disease progression and may be good candidates for primary PD prevention strategies (Belvisi et al. 2020). The brain is particularly vulnerable to oxidative damage because of its large oxygen consumption and lipid content. Typically, cells use antioxidant scavengers as a defense against ROS and free radicals. These scavengers include glutathione, ascorbic acid, vitamin E, carotenoids, flavonoids, polyphenols, and antioxidant enzymes. Ascorbic acid, or vitamin C, is found in high concentrations in the gray and white matter of the central nervous system. The brain and spinal cord also have the highest ascorbate concentrations of all body tissues. Dietary habits like increased antioxidant intake can therefore influence incidence of neurodegenerative disorders like PD (Esposito et al. 2002).

When autophagy becomes disrupted, cytotoxic/pathogenic molecules and cellular debris that are normally degraded by the lysosome can accumulate. An accumulation of protein, especially misfolded and pathogenic species, is a hallmark of several neurodegenerative diseases, including PD (Dickson et al. 2009). This clear connection has implicated autophagy pathways, specifically macroautophagy and CMA, as important processes in maintaining neuronal homeostasis and preventing the onset of PD. Naturally, therapeutics targeting defective autophagy pathways are of interest for treating patients with PD and preventing early-onset PD in high-risk individuals. Neurons exhibit high levels of autophagy relative to other cell types, as they are highly reliant on autophagic flux in order to dispatch protein aggregates and cellular debris (Son et al. 2012). As a result, any defect in the autophagy pathway will disproportionally affect neuronal cell types, where a subsequent increase in misfolded proteins can lead to the onset of neurodegenerative events. Under normal conditions, α-synuclein is found in the brain, largely localized to presynaptic terminals, where it functions in the regulation of synaptic vesicle trafficking (Vargas et al. 2014). Mutated, misfolded, and aggregated α-synuclein is degraded and removed from cells primarily by CMA but also by macroautophagy (Cuervo et al. 2004; Xilouri et al. 2009). In the case of PD, these pathogenic forms of α-synuclein aggregate in Lewy bodies inside neurons when autophagy is impaired (Vogiatzi et al. 2008). Accumulation of α-synuclein is directly associated with the death of dopaminergic neurons, the ultimate cause of PD progression (Stefanis 2012).

13.3.1 Genetics of Parkinson's Disease

13.3.1.1 α-Synuclein

Mutant α-synuclein is a common cause of PD pathology. Missense mutations in the membrane-binding domain and regulatory elements of the α-synuclein gene *SNCA* as well as duplications or triplications of the gene are the most frequent culprits of mutant α-synuclein accumulation in Lewy bodies (Polymeropoulos et al. 1997; Singleton et al. 2003). Posttranslational modifications (PTMs) of α-synuclein regulate its function and location and can determine its pathology. Phosphorylation at specific serine and tyrosine residues prevents α-synuclein aggregation and associated cytotoxic effects (Okochi et al. 2000; Ellis et al. 2001). Acetylation of α-synuclein at the N-terminus aids in preventing aggregation (Kang et al. 2012). At an unknown site near the N-terminus, α-synuclein is sumoylated by SUNO1 (Dorval and Fraser 2006). It remains to be determined if sumoylation primarily promotes or inhibits aggregation of α-synuclein as conflicting results have been observed (Krumova et al. 2011; Kim et al. 2011). α-Synuclein resides in Lewy bodies with ubiquitin and multiple ubiquitin ligases have been shown to ubiquitinate α-synuclein, including PRKN. It remains to be determined how ubiquitination of α-synuclein affects its role in PD pathology. In Lewy bodies, α-synuclein is commonly glycated which contributes to crosslinking and aggregation of α-synuclein (Mezey et al. 1998; Lansbury and Brice 2002; Münch et al. 2000). Glycosylation of α-synuclein may inhibit its aggregation and thus its cytotoxicity (Marotta et al. 2012). Tyrosine residues are sites of nitration in α-synuclein, and nitration of α-synuclein may contribute to its aggregation (Giasson et al. 2000). However, nitration has also been observed to prevent the formation of α-synuclein fibrils, making it another PTM with seemingly opposing effects on the contribution of α-synuclein to PD pathology. Truncation of α-synuclein is not associated with PD pathology. Lewy bodies contain mostly full-length α-synuclein, although the presence of C-terminally truncated α-synuclein increases the likelihood of the aggregation of full-length α-synuclein (Norris et al. 2003; Rochet et al. 2000).

Defective autophagy pathways that cannot clear α-synuclein aggregates contribute to PD pathology, but mutant α-synuclein can also inhibit autophagy on its own. In PD brains, CMA proteins have been seen to be decreased, and mutant α-synuclein also resists its own uptake by CMA (Cuervo et al. 2004; Alvarez-Erviti et al. 2010). Wild-type α-synuclein may resist uptake via macroautophagy, and its overexpression suppresses overall macroautophagic activity, allowing for not only the accumulation of α-synuclein but also of other PD-associated toxins and molecules. Inhibition of autophagy by α-synuclein overexpression can be rescued through both genetic and pharmacological methods (Winslow et al. 2010; Song et al. 2014). Mutant α-synuclein can inhibit mitophagy and removal of dysfunctional mitochondria (Takamura et al. 2011). α-Synuclein can also aggregate at the mitochondrial membrane, contributing to the role of mitochondrial toxicity and dysfunction in PD pathology (Hsu et al. 2000). In contrast to its vast array of anti-autophagic effects, there are various reports of α-synuclein overexpression-associated increase in autophagic flux and mitophagy. However, pro-autophagic activities induced by α-synuclein are not as well described (Xilouri et al. 2009; Choubey et al. 2011).

13.3.1.2 PINK1/PRKN

The autophagy genes PTEN-induced putative kinase 1 (*PINK1*) and *PRKN* (which encodes Parkin) are genes that have been implicated in inherited PD pathology. Mutations in PINK1 and PRKN are the most common cause of autosomal recessive early-onset PD (Kalia and Lang 2015; Corti et al. 2011). PINK1 participates in conjunction with the E3 ubiquitin ligase PARKIN to mediate mitophagy, a type of autophagy that selectively degrades mitochondria. PINK1 is found at mitochondria and PARKIN in the cytosol (Kondapalli et al. 2012; Martinez-Vicente 2017; Narendra et al. 2008). PINK1 surveys for damaged mitochondria and accumulates on their membranes and recruits PARKIN to the damaged mitochondria. PARKIN ubiquitinates the damaged mitochondria, signaling it for degradation in the lysosome via mitophagy. As previously described, α-synuclein can contribute to mitochondrial dysfunction, but interactions between PINK1-PARKIN and α-synuclein remain to be characterized (Geisler et al. 2010).

Disruption of complex I of the mitochondrial oxidative phosphorylation is a potential cause for PD via the overproduction of harmful ROS. Drug or chemically induced disruption, specifically by the mitochondrial toxin MPTP, has been linked to the PD-like symptoms of Parkinsonism, as well as sporadic PD. Mutations in mitochondrial DNA (mtDNA) also cause mitochondrial dysfunction. These mutations occur at higher frequencies in PD patients, and mutations in genes associated with replication and maintenance of mtDNA are associated with PD pathology (William Langston et al. 1983; Bender et al. 2006). A functional mitophagy pathway will traffic these dysfunctional mitochondria to the lysosome for degradation. However, mutations in the PINK1 and/or PARKIN that disrupt the mitophagy pathway fail to remove dysfunctional mitochondria from neurons. Mutations in PARKIN compromising its stability and solubility disrupt PINK1/PARKIN mitophagy induction and contribute to subsequent sporadic PD development. Loss-of-function mutations in PINK1 also contribute to inhibited mitophagy. Mutant PINK1 and PARKIN cause decreased turnover of mitochondrial respiratory chain elements and other mitochondria-associated proteins (Meng et al. 2011; Vincow et al. 2013). Mutations in PINK1 lead to defunct mitochondrial respiration and reduced ability to combat oxidative stress, as well as interruptions in mitochondrial membrane potential and energy production (Gautier et al. 2008; Diedrich et al. 2011). Most mutations in the *PRKN* gene interfere with its ability to ubiquitinate damaged mitochondria, while mutations in *PINK1* gene decrease its ability to interact with or activate PARKIN to the mitochondria (Lee et al. 2010; Puschmann et al. 2017). Damaged mitochondria that accumulate due to defective mitophagy produce harmful ROS and cause neuronal death. In this way, defects in mitophagy contribute to PD pathology.

13.3.1.3 LRRK2

The leucine-rich repeat kinase 2 (*LRRK2*) gene has been linked to both autophagy and inherited PD[49]. When overexpressed, LRRK2 induces macroautophagy and the formation of autophagosomes, and macroautophagy is reduced in cells when LRRK2 is silenced (Plowey et al. 2008; Bang et al. 2016). Conversely, LRRK2 has also been seen to play an anti-autophagic role where it inhibits autophagy of protein aggregates and silencing LRRK2 increases autophagy (Alegre-Abarrategui et al. 2009) . In line with its seemingly opposing functions, LRRK2 mutations in PD patient-derived fibroblasts have been associated with both increased macroautophagy and low levels of macroautophagy induction in response to stress (Bravo-San Pedro et al. 2013; Manzoni et al. 2013). Specifically, α-synuclein is seen to be increased and macroautophagy decreased in PD patient-derived neurons harboring LRRK2 mutations, which has been recapitulated in LRRK2 knockout models (Sánchez-Danés et al. 2012; Nguyen et al. 2011). The presence of mutant or excessive wild-type LRRK2 can inhibit CMA and cause α-synuclein aggregation in neurons (Yue and Yang 2013). LRRK2 has been specifically linked to lysosomal function via its kinase activity. In aged LRRK2 knockout mice, lysosomal proteins increase in the periphery, and inhibition of its kinase activity causes decreased localization of α-synuclein to the lysosome (Tong et al. 2012).

Similar to PINK1 and PARKIN, LRRK2 is important for mitophagy induction. Pathogenic mutant LRRK2 G2019S prevents the formation of the RHOT1/Miro1-LRRK2 complex, inhibiting mitophagy and the removal of harmful mitochondrial products (Hsieh et al. 2016). Reduced mitophagic activity and fewer autophagosomes were observed in fibroblasts harboring mutant LRRK2 due to decreased PINK1-PARKIN-mediated mitophagy (Korecka et al. 2019; Bonello et al. 2019) LRRK2 also interacts with PINK1-PARKIN mitophagy through RAB10, which signals optineurin to localize to depolarized mitochondria (Wauters et al. 2020). Conversely, this same mutant LRRK2 has been shown to stimulate mitophagy by interacting with the autophagy regulator Unc51-like kinase 1 (ULK1) and activation of the ERK1/2 pathway (Zhu et al. 2013). As mentioned previously, decreased mitophagy results in accumulation of dysfunctional mitochondria which contribute to dopaminergic neuron death and PD pathology.

13.3.1.4 DJ-1

DJ-1 is primarily known to serve a protective role against oxidative stress but has also been implicated in autophagy regulation as well as both sporadic and autosomal recessive PD pathology (Son et al. 2012). The oxidized, neuroprotective form of DJ-1 has been found in high levels in PD brains, but a reduction of oxidized DJ-1 has been observed in the brains of sporadic PD patients. Mutant L166P DJ-1 fails to dimerize, stalling lysosomal activity causing mitochondrial health to suffer. DJ-1 functions in the activation of mitophagy and removal of dysfunctional mitochondria and the direct removal of harmful ROS. DJ-1 responds to excessive ROS production by localizing to the dysfunctional mitochondria (Junn et al. 2009; Hayashi et al. 2009). Binding of DJ-1 to complex I occurs under normal conditions but is increased during oxidative stress. DJ-1 deficiency results in compromised mitochondrial membrane potential and the failure to activate mitophagy. DJ-1 acts in tandem with PINK1 and PRKN and exhibits similar neuroprotective effects against rotenone to PINK1. However, under nitrosative stress, DJ-1 nitrosylates and inhibits PTEN, activating the Akt signaling pathway (Thomas et al. 2011; Numajiri et al. 2011).

DJ-1 regulates CMA, mediates removal of α-synuclein aggregates, and even inhibits initial aggregation of α-synuclein (Martinat et al. 2004). In various neuronal cell types, mutation, knockdown, and deficiency of DJ-1 inhibit autophagic removal of α-synuclein and allow excessive α-synuclein aggregation, causing subsequent neurodegeneration. Deficiency in DJ-1 results in increased lysosomal LAMP2A degradation and reduced CMA-mediated degradation of α-synuclein (Xu et al. 2017). In agreement with this, overexpression of DJ-1 enhances autophagic clearing of the PD toxin rotenone and limits α-synuclein aggregation, protecting against neuron death (Martinat et al. 2004; Xu et al. 2017).

13.3.2 Nutrigenomics and Parkinson's Disease

13.3.2.1 Dietary Influences on PD

Dietary fat intake can potentially affect PD risk because polyunsaturated fatty acids (PUFA) may contribute to oxidative stress and neuroinflammation. PUFA is found in high concentrations in neural membranes and can be a source of oxygen radicals through lipid peroxidation, which may cause mitochondrial dysfunction and contribute to PD. PUFA may also promote α-synuclein oligomerization and aggregation in DA neurogenic cells. Most membrane PUFA are synthesized from $n - 6$ linoleic and $n - 3$ linoleic acids, which are obtained from the diet and are precursors for synthesis of long-chain PUFA. $n - 6$ PUFA are precursors for proinflammatory prostaglandins, so it is biologically plausible that high $n - 6$ PUFA intake is associated with higher PD risk. Additionally, there is a borderline inverse association between PD and arachidonic acid, meaning a dietary intake of arachidonic acid may be a protective factor for PD (Perez-Pardo et al. 2019). There is evidence that caffeine intake in coffee and tea exerts a protective effect against the development of PD and reduces the risk of developing PD by half. This inverse association is more pronounced in men than in women (Belvisi et al. 2020) . Caffeine is a stimulatory drug metabolized by the CYPIA2 isoenzyme of the p450 family which impacts the central nervous system (Ishihara and Brayne 2005). The protective effects against PD are plausibly mediated by blocking adenosine A2A receptors. Caffeine protects against neurotoxicity and degeneration of the DA nigrostriatal system by preventing apoptosis. Caffeine has also been shown to protect against blood-brain barrier dysfunction in animal models, preventing DA loss. Tea also contains flavonoids like (−)-epigallocatechin-3-gallate (EGCG) found in green tea, which exert neuroprotective activity via microglial inhibition and blocking of DA transporters (Di Giovanni 2009).

Dairy products are the only food group that has been consistently linked with higher incidence of PD, although this is not due to the calcium, vitamin D, or fat concentrations in these products. Higher dairy food consumption has been shown to reduce uric acid levels, which may be neuroprotective against oxidative damage from ROS and reactive nitrogen species (RNS). Higher plasma levels of uric acid have been linked to lower incidence of PD (Di Giovanni 2009). In addition, urate may be the first molecular predictor of clinical progression in PD, and higher urate levels may be able to reduce disease progression (Boulos et al. 2019). The Mediterranean pattern diet, which incorporates many plant foods, fresh fruit, olive oil, and fish and poultry and low amounts of red meat, has been shown to be protective against PD. On the other hand, a western pattern of eating, which includes high intakes of red meat, sugary drinks and desserts, high fat dairy products, eggs, and refined grains, has been shown to increase PD risk (Di Giovanni 2009).

13.3.2.2 Nutrition-Based Therapeutic Strategies in PD

Accumulation of α-synuclein is the pathological hallmark of PD and is directly associated with neuronal death and onset of PD (Dickson et al. 2009; Stefanis 2012). Since dysfunctional autophagy pathways fail to clear α-synuclein, it is then an intuitive conclusion to target restoration of dysfunctional autophagy and mitophagy pathways in order to treat PD. Traditionally, PD treatments have focused on mitigating the symptoms and limiting neuron death, while dopamine supplementation via L-Dopa helps restore some lost motor function (Lewitt 2015). Inhibition of the mTOR pathway using mTOR inhibitors, including rapamycin, has been found to increase autophagy and aggregated protein degradation via the lysosome in animal models. This restoration of autophagy is accompanied by relief of PD symptoms and decreased neurodegeneration (Liu et al. 2013; Dehay et al. 2010). Treatment with rapamycin results in many off-target effects and is not ideal outside of serious PD cases (Bové et al. 2011). Consequently, natural products that induce autophagy independently of mTOR are of recent interest as therapeutics for PD. Many common foods, most notably fruits and vegetables, are abundant with various phenolic compounds, which are widely known for their anti-inflammatory and antioxidant effects (Recio et al. 2012). Recently, phenolic compounds were seen to induce autophagy and even alleviate symptoms in induced PD animal models (Ferretta et al. 2014; Guo et al. 2016) . Another potential therapeutic strategy for PD treatment includes small molecule drugs targeting lysosomal activation, including ambroxol and isofagomine (Richter et al. 2014; McNeill et al. 2014). Despite the range of novel therapeutics on the horizon, there is no known cure for PD. It remains a complex disease that severely diminishes the quality of life of a significant proportion of the elderly population and a high priority for innovative therapeutic development.

13.4 Conclusions

Dietary modifications to increase autophagy in neurodegenerative disease represent a promising therapeutic approach to treat multifactorial disorders such as AD and PD. In the future, disease-modifying therapeutic targets may be achieved through precision medicine approaches that target nutritional deficiencies or imbalances to reduce the burden of neurodegenerative disease in aging populations.

References

Akomolafe A, Beiser A, Meigs JB, Au R, Green RC, Farrer LA, Wolf PA, Seshadri S (2006) Diabetes mellitus and risk of developing Alzheimer disease: results from the Framingham study. Arch Neurol 63:1551–1555

Alavi Naini SM, Soussi-Yanicostas N (2015) Tau hyperphosphorylation and oxidative stress, a critical vicious circle in neurodegenerative tauopathies? Oxidative Med Cell Longev 2015: 151979. https://doi.org/10.1155/2015/151979

Alegre-Abarrategui J, Christian H, Lufino MMP, Mutihac R, Venda LL, Ansorge O, Wade-Martins R (2009) LRRK2 regulates autophagic activity and localizes to specific membrane microdomains in a novel human genomic reporter cellular model. Hum Mol Genet 18:4022–4034

Alvarez-Erviti L, Rodriguez-Oroz MC, Cooper JM, Caballero C, Ferrer I, Obeso JA, Schapira AHV (2010) Chaperone-mediated autophagy markers in Parkinson disease brains. Arch Neurol 67: 1464–1472

Armstrong RA (2019) Risk factors for Alzheimer's disease. Folia Neuropathol 57:87–105

Arvanitakis Z, Wilson RS, Bienias JL, Evans DA, Bennett DA (2004) Diabetes mellitus and risk of alzheimer disease and decline in cognitive function. Arch Neurol 61(5):661–666

Arvanitakis Z, Schneider JA, Wilson RS, Bienias JL, Kelly JF, Evans DA, Bennett DA (2008) Statins, incident Alzheimer disease, change in cognitive function, and neuropathology. Neurology 70:1795–1802

Bang Y, Kim KS, Seol W, Choi HJ (2016) LRRK2 interferes with aggresome formation for autophagic clearance. Mol Cell Neurosci 75:71–80

Bateman GA, Levi CR, Schofield P, Wang Y, Lovett EC (2006) Quantitative measurement of cerebral haemodynamics in early vascular dementia and Alzheimer's disease. J Clin Neurosci 13:563–568

Beal MF (2002) Oxidatively modified proteins in aging and disease. Free Radic Biol Med 32:797–803

Begg DP (2015) Insulin transport into the brain and cerebrospinal fluid. In: Vitamins and hormones. Academic, New York, pp 229–248

Belinson H, Lev D, Masliah E, Michaelson DM (2008) Activation of the amyloid cascade in apolipoprotein E4 transgenic mice induces lysosomal activation and neurodegeneration resulting in marked cognitive deficits. J Neurosci 28:4690–4701

Belvisi D, Pellicciari R, Fabbrini G, Tinazzi M, Berardelli A, Defazio G (2020) Modifiable risk and protective factors in disease development, progression and clinical subtypes of Parkinson's disease: what do prospective studies suggest? Neurobiol Dis 134:104671. https://doi.org/10.1016/j.nbd.2019.104671

Bender A, Krishnan KJ, Morris CM et al (2006) High levels of mitochondrial DNA deletions in substantia nigra neurons in aging and Parkinson disease. Nat Genet 38:515–517

Beyer K, Lao JI, Latorre P, Ariza A (2005) Age at onset: an essential variable for the definition of genetic risk factors for sporadic Alzheimer's disease. Ann N Y Acad Sci 1057:260–278

Boland B, Kumar A, Lee S, Platt FM, Wegiel J, Yu WH, Nixon RA (2008) Autophagy induction and autophagosome clearance in neurons: relationship to autophagic pathology in Alzheimer's disease. J Neurosci 28:6926–6937

Bondolfi L, Ermini F, Long JM, Ingram DK, Jucker M (2004) Impact of age and caloric restriction on neurogenesis in the dentate gyrus of C57BL/6 mice. Neurobiol Aging 25:333–340

Bonello F, Hassoun SM, Mouton-Liger F et al (2019) LRRK2 impairs PINK1/Parkin-dependent mitophagy via its kinase activity: pathologic insights into Parkinson's disease. Hum Mol Genet 28:1645–1660

Boulos C, Yaghi N, El HR, Heraoui GNHA, Fakhoury-Sayegh N (2019) Nutritional risk factors, microbiota and parkinson's disease: what is the current evidence? Nutrients 11(8):1896. https://doi.org/10.3390/nu11081896

Bové J, Martínez-Vicente M, Vila M (2011) Fighting neurodegeneration with rapamycin: mechanistic insights. Nat Rev Neurosci 12:437–452

Bravo-San Pedro JM, Niso-Santano M, Gómez-Sánchez R et al (2013) The LRRK2 G2019S mutant exacerbates basal autophagy through activation of the MEK/ERK pathway. Cell Mol Life Sci 70:121–136

Burillo J, Marqués P, Jiménez B, González-Blanco C, Benito M, Guillén C (2021) Insulin resistance and diabetes mellitus in Alzheimer's disease. Cell 10:1–42

Busche MA, Hyman BT (2020) Synergy between amyloid-β and tau in Alzheimer's disease. Nat Neurosci 23:1183–1193

Butterfield DA, Kanski J (2001) Brain protein oxidation in age-related neurodegenerative disorders that are associated with aggregated proteins. Mech Ageing Dev 122:945–962

Butterfield DA, Swomley AM, Sultana R (2013) Amyloid β-peptide (1-42)-induced oxidative stress in Alzheimer disease: importance in disease pathogenesis and progression. Antioxidants Redox Signal 19:823–835

Caccamo A, De Pinto V, Messina A, Branca C, Oddo S (2014) Genetic reduction of mammalian target of rapamycin ameliorates Alzheimer's disease-like cognitive and pathological deficits by restoring hippocampal gene expression signature. J Neurosci 34:7988–7998

Candeias E, Duarte AI, Sebastião I et al (2017) Middle-aged diabetic females and males present distinct susceptibility to Alzheimer disease-like pathology. Mol Neurobiol 54:6471–6489

Cerri S, Blandini F (2019) Role of autophagy in Parkinson's disease. Curr Med Chem 26:3702–3718

Cha MY, Kim DK, Mook-Jung I (2015) The role of mitochondrial DNA mutation on neurodegenerative diseases. Exp Mol Med 47:e150

Chandra S, Jana M, Pahan K (2018) Aspirin induces lysosomal biogenesis and attenuates amyloid plaque pathology in a mouse model of Alzheimer's disease via PPARα. J Neurosci 38:6682–6699

Chaudhuri TK, Paul S (2006) Protein-misfolding diseases and chaperone-based therapeutic approaches. FEBS J 273:1331–1349

Chen Z, Zhong C (2014) Oxidative stress in Alzheimer's disease. Neurosci Bull 30:271–281

Cho MH, Cho K, Kang HJ, Jeon EY, Kim HS, Kwon HJ, Kim HM, Kim DH, Yoon SY (2014) Autophagy in microglia degrades extracellular β-amyloid fibrils and regulates the NLRP3 inflammasome. Autophagy 10:1761–1775

Chong CM, Ke M, Tan Y, Huang Z, Zhang K, Ai N, Ge W, Qin D, Lu JH, Su H (2018) Presenilin 1 deficiency suppresses autophagy in human neural stem cells through reducing γ-secretase-independent ERK/CREB signaling. Cell Death Dis 9(9):879. https://doi.org/10.1038/s41419-018-0945-7

Choubey V, Safiulina D, Vaarmann A, Cagalinec M, Wareski P, Kuum M, Zharkovsky A, Kaasik A (2011) Mutant A53T α-Synuclein induces neuronal death by increasing mitochondrial autophagy. J Biol Chem 286:10814–10824

Congdon EE, Wu JW, Myeku N, Figueroa YH, Herman M, Marinec PS, Gestwicki JE, Dickey CA, Yu WH, Duff KE (2012) Methylthioninium chloride (methylene blue) induces autophagy and attenuates tauopathy in vitro and in vivo. Autophagy 8:609–622

Cordero JG, García-Escudero R, Avila J, Gargini R, García-Escudero V (2018) Benefit of oleuropein aglycone for Alzheimer's disease by promoting autophagy. Oxidative Med Cell Longev 2018:5010741. https://doi.org/10.1155/2018/5010741
Corti O, Lesage S, Brice A (2011) What genetics tells us about the causes and mechanisms of Parkinson's disease. Physiol Rev 91:1161–1218
Costa LG, Garrick JM, Roquè PJ, Pellacani C (2016) Mechanisms of neuroprotection by quercetin: counteracting oxidative stress and more. Oxidative Med Cell Longev 2016:2986796. https://doi.org/10.1155/2016/2986796
Cuervo AM, Stafanis L, Fredenburg R, Lansbury PT, Sulzer D (2004) Impaired degradation of mutant α-synuclein by chaperone-mediated autophagy. Science 805:1292–1295
De Mello NP, Orellana AM, Mazucanti CH, De Morais LG, Scavone C, Kawamoto EM (2019) Insulin and autophagy in neurodegeneration. Front Neurosci 13:1–17
Dehay B, Bové J, Rodríguez-Muela N, Perier C, Recasens A, Boya P, Vila M (2010) Pathogenic lysosomal depletion in Parkinson's disease. J Neurosci 30:12535–12544
Deng H, Mi M-T (2016) Resveratrol attenuates Aβ 25–35 caused neurotoxicity by inducing autophagy through the TyrRS-PARP1-SIRT1 signaling pathway. Neurochem Res 41:2367–2379
Di Giovanni G (2009) A diet for dopaminergic neurons? In: Birth, life death dopaminergic neurons substantia nigra. Springer, Vienna, pp 317–331
Di Meco A, Curtis ME, Lauretti E, Praticò D (2020) Autophagy dysfunction in Alzheimer's disease: mechanistic insights and new therapeutic opportunities. Biol Psychiatry 87:797–807
Dickson DW, Braak H, Duda JE et al (2009) Neuropathological assessment of Parkinson's disease: refining the diagnostic criteria. Lancet Neurol 8:1150–1157
Diedrich M, Kitada T, Nebrich G, Koppelstaetter A, Shen J, Zabel C, Klose J, Mao L (2011) Brain region specific mitophagy capacity could contribute to selective neuronal vulnerability in Parkinson's disease. Proteome Sci 9:59
Dionísio PA, Amaral JD, Rodrigues CMP (2021) Oxidative stress and regulated cell death in Parkinson's disease. Ageing Res Rev 67:101263. https://doi.org/10.1016/j.arr.2021.101263
Dorval V, Fraser PE (2006) Small ubiquitin-like modifier (SUMO) modification of natively unfolded proteins tau and α-synuclein. J Biol Chem 281:9919–9924
Du J, Liang Y, Xu F, Sun B, Wang Z (2013) Trehalose rescues Alzheimer's disease phenotypes in APP/PS1 transgenic mice. J Pharm Pharmacol 65:1753–1756
Ellis CE, Schwartzberg PL, Grider TL, Fink DW, Nussbaum RL (2001) α-Synuclein is phosphorylated by members of the Src family of protein-tyrosine kinases. J Biol Chem 276:3879–3884
Eshraghi M, Adlimoghaddam A, Mahmoodzadeh A, Sharifzad F, Yasavoli-sharahi H, Lorzadeh S, Albensi BC, Ghavami S (2021) Alzheimer's disease pathogenesis: role of autophagy and mitophagy focusing in microglia. Int J Mol Sci 22:1–36
Esposito E, Rotilio D, Di Matteo V, Di Giulio C, Cacchio M, Algeri S (2002) A review of specific dietary antioxidants and the effects on biochemical mechanisms related to neurodegenerative processes. Neurobiol Aging 23:719–735
Fernández-Sanz P, Ruiz-Gabarre D, García-Escudero V (2019) Modulating effect of diet on Alzheimer's disease. Diseases 7:12
Ferretta A, Gaballo A, Tanzarella P et al (2014) Effect of resveratrol on mitochondrial function: implications in parkin-associated familiar Parkinson's disease. Biochim Biophys Acta Mol basis Dis 1842:902–915
Galasko DR, Peskind E, Clark CM et al (2012) Antioxidants for Alzheimer disease a randomized clinical trial with cerebrospinal fluid biomarker measures. Trial Regist Clin Identifier NCT00117403. Arch Neurol 69:836–841
Gautier CA, Kitada T, Shen J (2008) Loss of PINK1 causes mitochondrial functional defects and increased sensitivity to oxidative stress. Proc Natl Acad Sci U S A 105:11364–11369
Geisler S, Holmström KM, Skujat D, Fiesel FC, Rothfuss OC, Kahle PJ, Springer W (2010) PINK1/Parkin-mediated mitophagy is dependent on VDAC1 and p62/SQSTM1. Nat Cell Biol 12:119–131

Ghadernezhad N, Khalaj L, Pazoki-Toroudi H, Mirmasoumi M, Ashabi G (2016) Metformin pretreatment enhanced learning and memory in cerebral forebrain ischaemia: the role of the AMPK/BDNF/P70SK signalling pathway. Pharm Biol 54:2211–2219

Giasson BI, Duda JE, Murray IVJ, Chen Q, Souza JM, Hurtig HI, Ischiropoulos H, Trojanowski JQ, Lee VMY (2000) Oxidative damage linked to neurodegeneration by selective α-synuclein nitration in synucleinopathy lesions. Science 290:985–989

Grote CW, Wright DE (2016) A role for insulin in diabetic neuropathy. Front Neurosci 10:581

Guo YJ, Dong SY, Cui XX, Feng Y, Liu T, Yin M, Kuo SH, Tan EK, Zhao WJ, Wu YC (2016) Resveratrol alleviates MPTP-induced motor impairments and pathological changes by autophagic degradation of α-synuclein via SIRT1-deacetylated LC3. Mol Nutr Food Res 60: 2161–2175

Hara T, Nakamura K, Matsui M et al (2006) Suppression of basal autophagy in neural cells causes neurodegenerative disease in mice. Nature 441:885–889

Hayashi T, Ishimori C, Takahashi-Niki K, Taira T, Kim Y-c, Maita H, Maita C, Ariga H, Iguchi-Ariga SMM (2009) DJ-1 binds to mitochondrial complex I and maintains its activity. Biochem Biophys Res Commun 390:667–672

Heckmann BL, Teubner BJW, Tummers B, Guy CS, Zakharenko SS, Green DR (2019) LC3-associated endocytosis facilitates β-amyloid clearance and mitigates neurodegeneration in murine Alzheimer's disease. Cell 178:536–551

Hochrainer K, Jackman K, Anrather J, Iadecola C (2012) Reperfusion rather than ischemia drives the formation of ubiquitin aggregates after middle cerebral artery occlusion. Stroke 43:2229–2235

Horton AC, Ehlers MD (2003) Neuronal polarity and trafficking. Neuron 40:277–295

Houtman J, Freitag K, Gimber N, Schmoranzer J, Heppner FL, Jendrach M (2019) Beclin1-driven autophagy modulates the inflammatory response of microglia via NLRP 3. EMBO J 38(4): e99430. https://doi.org/10.15252/embj.201899430

Hsieh CH, Shaltouki A, Gonzalez AE, Bettencourt da Cruz A, Burbulla LF, St Lawrence E, Schüle B, Krainc D, Palmer TD, Wang X (2016) Functional impairment in miro degradation and mitophagy is a shared feature in familial and sporadic Parkinson's disease. Cell Stem Cell 19:709–724

Hsu LJ, Sagara Y, Arroyo A, Rockenstein E, Sisk A, Mallory M, Wong J, Takenouchi T, Hashimoto M, Masliah E (2000) alpha-synuclein promotes mitochondrial deficit and oxidative stress. Am J Pathol 157:401–410

Hu BR, Martone ME, Jones YZ, Liu CL (2000) Protein aggregation after transient cerebral ischemia. J Neurosci 20:3191–3199

Huang Y, Mahley RW (2014) Apolipoprotein E: structure and function in lipid metabolism, neurobiology, and Alzheimer's diseases. Neurobiol Dis 72:3–12

Ishihara L, Brayne C (2005) A systematic review of nutritional risk factors of Parkinson's disease. Nutr Res Rev 18:259–282

Ji ZS, Müllendorff K, Cheng IH, Miranda RD, Huang Y, Mahley RW (2006) Reactivity of apolipoprotein E4 and amyloid β peptide: lysosomal stability and neurodegeneration. J Biol Chem 281:2683–2692

Junn E, Jang WH, Zhao X, Jeong BS, Mouradian MM (2009) Mitochondrial localization of DJ-1 leads to enhanced neuroprotection. J Neurosci Res 87:123–129

Kahl A, Blanco I, Jackman K, Baskar J, Milaganur Mohan H, Rodney-Sandy R, Zhang S, Iadecola C, Hochrainer K (2018) Cerebral ischemia induces the aggregation of proteins linked to neurodegenerative diseases. Sci Rep 8(1):2701. https://doi.org/10.1038/s41598-018-21063-z

Kalia LV, Lang AE (2015) Parkinson's disease. Lancet 386:896–912

Kang L, Moriarty GM, Woods LA, Ashcroft AE, Radford SE, Baum J (2012) N-terminal acetylation of α-synuclein induces increased transient helical propensity and decreased aggregation rates in the intrinsically disordered monomer. Protein Sci 21:911–917

Karuppagounder SS, Xu H, Shi Q, Chen LH, Pedrini S, Pechman D, Baker H, Beal MF, Gandy SE, Gibson GE (2009) Thiamine deficiency induces oxidative stress and exacerbates the plaque pathology in Alzheimer's mouse model. Neurobiol Aging 30:1587–1600

Katila N, Bhurtel S, Shadfar S, Srivastav S, Neupane S, Ojha U, Jeong GS, Choi DY (2017) Metformin lowers α-synuclein phosphorylation and upregulates neurotrophic factor in the MPTP mouse model of Parkinson's disease. Neuropharmacology 125:396–407

Kim J, Basak JM, Holtzman DM (2009) The role of apolipoprotein E in Alzheimer's disease. Neuron 63:287–303

Kim YM, Jang WH, Quezado MM, Oh Y, Chung KC, Junn E, Mouradian MM (2011) Proteasome inhibition induces α-synuclein SUMOylation and aggregate formation. J Neurol Sci 307:157–161

Kim J, Yoon H, Basak J, Kim J (2014) Apolipoprotein E in synaptic plasticity and Alzheimer's disease: potential cellular and molecular mechanisms. Mol Cells 37:833–840

Kloske CM, Wilcock DM (2020) The important Interface between apolipoprotein E and Neuroinflammation in Alzheimer's disease. Front Immunol 11:754. https://doi.org/10.3389/fimmu.2020.00754

Komatsu M, Waguri S, Chiba T et al (2006) Loss of autophagy in the central nervous system causes neurodegeneration in mice. Nature 441:880–884

Kondapalli C, Kazlauskaite A, Zhang N et al (2012) PINK1 is activated by mitochondrial membrane potential depolarization and stimulates Parkin E3 ligase activity by phosphorylating serine 65. Open Biol 2:120080

Korecka JA, Thomas R, Christensen DP, Hinrich AJ, Ferrari EJ, Levy SA, Hastings ML, Hallett PJ, Isacson O (2019) Mitochondrial clearance and maturation of autophagosomes are compromised in LRRK2 G2019S familial Parkinson's disease patient fibroblasts. Hum Mol Genet 28:3232–3243

Krumova P, Meulmeester E, Garrido M et al (2011) Sumoylation inhibits α-synuclein aggregation and toxicity. J Cell Biol 194:49–60

Labbadia J, Morimoto RI (2015) The biology of proteostasis in aging and disease. Annu Rev Biochem 84:435–464

Lahiri DK, Sambamurti K, Bennett DA (2004) Apolipoprotein gene and its interaction with the environmentally driven risk factors: molecular, genetic and epidemiological studies of Alzheimer's disease. Neurobiol Aging 25:651–660

Lansbury PT, Brice A (2002) Genetics of Parkinson's disease and biochemical studies of implicated gene products: commentary. Curr Opin Cell Biol 14:653–660

Lapchak PA, Araujo DM (2001) Preclinical development of neurosteroids as neuroprotective agents for the treatment of neurodegenerative diseases. Int Rev Neurobiol 46:379–397

Lee JY, Nagano Y, Taylor JP, Lim KL, Yao TP (2010) Disease-causing mutations in Parkin impair mitochondrial ubiquitination, aggregation, and HDAC6-dependent mitophagy. J Cell Biol 189:671–679

Lewitt PA (2015) Levodopa therapy for Parkinson's disease: pharmacokinetics and pharmacodynamics. Mov Disord 30:64–72

Li Q, Liu Y, Sun M (2017) Autophagy and Alzheimer's disease. Cell Mol Neurobiol 37:377–388

Li HH, Lin CL, Huang CN (2018) Neuroprotective effects of statins against amyloid β-induced neurotoxicity. Neural Regen Res 13:198–206

Liao F, Yoon H, Kim J (2017) Apolipoprotein E metabolism and functions in brain and its role in Alzheimer's disease. Curr Opin Lipidol 28:60–67

Limanaqi F, Biagioni F, Gambardella S, Familiari P, Frati A, Fornai F (2020) Promiscuous roles of autophagy and proteasome in neurodegenerative proteinopathies. Int J Mol Sci 21(8):3028. https://doi.org/10.3390/ijms21083028

Liu J, Li L (2019) Targeting autophagy for the treatment of Alzheimer's disease: challenges and opportunities. Front Mol Neurosci 12:1–9

Liu K, Shi N, Sun Y, Zhang T, Sun X (2013) Therapeutic effects of rapamycin on MPTP-induced parkinsonism in mice. Neurochem Res 38:201–207

Liu Z, Li T, Li P, Wei N, Zhao Z, Liang H, Ji X, Chen W, Xue M, Wei J (2015) The ambiguous relationship of oxidative stress, tau hyperphosphorylation, and autophagy dysfunction in Alzheimer's disease. Oxidative Med Cell Longev 2015:352723. https://doi.org/10.1155/2015/352723

Liu R, Yang J, Liu L, Lu Z, Shi Z, Ji W, Shen J, Zhang X (2020) An "amyloid-β cleaner" for the treatment of Alzheimer's disease by normalizing microglial dysfunction. Adv Sci 7(2):1901555. https://doi.org/10.1002/advs.201901555

Long JM, Holtzman DM (2019) Alzheimer disease: an update on pathobiology and treatment strategies. Cell 179:312–339

Lovell MA, Ehmann WD, Butler SM, Markesbery WR (1995) Elevated thiobarbituric acid-reactive substances and antioxidant enzyme activity in the brain in Alzheimer's disease. Neurology 45: 1594–1601

Lu M, Su C, Qiao C, Bian Y, Ding J, Hu G (2016) Metformin prevents dopaminergic neuron death in MPTP/P-induced mouse model of Parkinson's disease via autophagy and mitochondrial ROS clearance. Int J Neuropsychopharmacol 19:1–11

Lucin KM, O'Brien CE, Bieri G et al (2013) Microglial beclin 1 regulates retromer trafficking and phagocytosis and is impaired in Alzheimer's disease. Neuron 79:873–886

MacKnight C, Rockwood K, Awalt E, McDowell I (2002) Diabetes mellitus and the risk of dementia, Alzheimer's disease and vascular cognitive impairment in the Canadian study of health and aging. Dement Geriatr Cogn Disord 14:77–83

Manzoni C, Mamais A, Dihanich S et al (2013) Pathogenic Parkinson's disease mutations across the functional domains of LRRK2 alter the autophagic/lysosomal response to starvation. Biochem Biophys Res Commun 441:862–866

Marcus DL, Thomas C, Rodriguez C, Simberkoff K, Tsai JS, Strafaci JA, Freedman ML (1998) Increased peroxidation and reduced antioxidant enzyme activity in Alzheimer's disease. Exp Neurol 150:40–44

Margolis RU, Altszuler N (1967) Insulin in the cerebrospinal fluid. Nature 215:1375–1376

Marotta NP, Cherwien CA, Abeywardana T, Pratt MR (2012) O-GlcNAc modification prevents peptide-dependent acceleration of α-synuclein aggregation. Chembiochem 13:2665–2670

Martin B, Pearson M, Kebejian L et al (2007) Sex-dependent metabolic, neuroendocrine, and cognitive responses to dietary energy restriction and excess. Endocrinology 148:4318–4333

Martinat C, Shendelman S, Jonason A, Leete T, Beal MF, Yang L, Floss T, Abeliovich A (2004) Sensitivity to oxidative stress in DJ-1-deficient dopamine neurons: an ES-derived cell model of primary parkinsonism. PLoS Biol 2:e327

Martinez-Vicente M (2017) Neuronal mitophagy in neurodegenerative diseases. Front Mol Neurosci 10:64. https://doi.org/10.3389/fnmol.2017.00064

Masters CL, Bateman R, Blennow K, Rowe CC, Sperling RA, Cummings JL (2015) Alzheimer's disease. Nat Rev Dis Prim 1:1–18

Mazucanti CH, Liu QR, Lang D, Huang N, O'Connell JF, Camandola S, Egan JM (2019) Release of insulin produced by the choroids plexus is regulated by serotonergic signaling. JCI Insight 4(23):e131682. https://doi.org/10.1172/jci.insight.131682

McNeill A, Magalhaes J, Shen C et al (2014) Ambroxol improves lysosomal biochemistry in glucocerebrosidase mutation-linked Parkinson disease cells. Brain 137:1481–1495

Meng F, Yao D, Shi Y et al (2011) Oxidation of the cysteine-rich regions of parkin perturbs its E3 ligase activity and contributes to protein aggregation. Mol Neurodegener 6:34

Mezey E, Dehejia AM, Harta G, Suchy SF, Nussbaum RL, Brownstein MJ, Polymeropoulos MH (1998) Alpha synuclein is present in Lewy bodies in sporadic Parkinson's disease. Mol Psychiatry 3:493–499

Mizushima N (2005) Aβ generation in autophagic vacuoles. J Cell Biol 171:15–17

Mizushima N, Komatsu M (2011) Autophagy: renovation of cells and tissues. Cell 147:728–741

Münch G, Lüth HJ, Wong A, Arendt T, Hirsch E, Ravid R, Riederer P (2000) Crosslinking of α-synuclein by advanced glycation endproducts—an early pathophysiological step in Lewy body formation? J Chem Neuroanat 20:253–257

Spilman P, Podlutskaya N, Hart MJ, Debnath J, Gorostiza O, Bredesen D, Richardson A, Strong R, Galvan V (2010) Inhibition of mTOR by rapamycin abolishes cognitive deficits and reduces amyloid-β levels in a mouse model of Alzheimer's disease. PLoS One 5(4):e9979. https://doi.org/10.1371/journal.pone.0009979

Steen E, Terry BM, Rivera EJ, Cannon JL, Neely TR, Tavares R, Xu XJ, Wands JR, De La Monte SM (2005) Impaired insulin and insulin-like growth factor expression and signaling mechanisms in Alzheimer's disease—is this type 3 diabetes? J Alzheimers Dis 7:63–80

Stefanis L (2012) α-Synuclein in Parkinson's disease. Cold Spring Harb Perspect Med 2:a009399

Strittmatter WJ, Roses AD (1996) Apolipoprotein E and Alzheimer's disease. Annu Rev Neurosci 19:53–77

Sun X-Y, Dong Q-X, Zhu J, Sun X, Zhang L-F, Qiu M, Yu X-L, Liu R-T (2019) Resveratrol rescues tau-induced cognitive deficits and neuropathology in a mouse model of tauopathy. Curr Alzheimer Res 16:710–722

Takamura A, Komatsu M, Hara T, Sakamoto A, Kishi C, Waguri S, Eishi Y, Hino O, Tanaka K, Mizushima N (2011) Autophagy-deficient mice develop multiple liver tumors. Genes Dev 25: 795–800

Takano K, Koarashi K, Kawabe K, Itakura M, Nakajima H, Moriyama M, Nakamura Y (2018) Insulin expression in cultured astrocytes and the decrease by amyloid β. Neurochem Int 119: 171–177

Thomas KJ, McCoy MK, Blackinton J, Beilina A, van der Brug M, Sandebring A, Miller D, Maric D, Cedazo-Minguez A, Cookson MR (2011) DJ-1 acts in parallel to the PINK1/parkin pathway to control mitochondrial function and autophagy. Hum Mol Genet 20:40–50

Tong Y, Giaime E, Yamaguchi H, Ichimura T, Liu Y, Si H, Cai H, Bonventre JV, Shen J (2012) Loss of leucine-rich repeat kinase 2 causes age-dependent bi-phasic alterations of the autophagy pathway. Mol Neurodegener 7:2

Van Acker ZP, Bretou M, Annaert W (2019) Endo-lysosomal dysregulations and late-onset Alzheimer's disease: impact of genetic risk factors. Mol Neurodegener 14(1):20. https://doi.org/10.1186/s13024-019-0323-7

Vargas KJ, Makani S, Davis T, Westphal CH, Castillo PE, Chandra SS (2014) Synucleins regulate the kinetics of synaptic vesicle endocytosis. J Neurosci 34:9364–9376

Vincow ES, Merrihew G, Thomas RE, Shulman NJ, Beyer RP, MacCoss MJ, Pallanck LJ (2013) The PINK1-Parkin pathway promotes both mitophagy and selective respiratory chain turnover in vivo. Proc Natl Acad Sci U S A 110:6400–6405

Vingtdeux V, Giliberto L, Zhao H et al (2010) AMP-activated protein kinase signaling activation by resveratrol modulates amyloid-β peptide metabolism. J Biol Chem 285:9100–9113

Vogiatzi T, Xilouri M, Vekrellis K, Stefanis L (2008) Wild type α-synuclein is degraded by chaperone-mediated autophagy and macroautophagy in neuronal cells. J Biol Chem 283: 23542–23556

Wang D, Chan CC, Cherry S, Hiesinger PR (2013) Membrane trafficking in neuronal maintenance and degeneration. Cell Mol Life Sci 70:2919–2934

Wang C, Zhang X, Teng Z, Zhang T, Li Y (2014) Downregulation of PI3K/Akt/mTOR signaling pathway in curcumin-induced autophagy in APP/PS1 double transgenic mice. Eur J Pharmacol 740:312–320

Watanabe Y, Taguchi K, Tanaka M (2020) Ubiquitin, autophagy and neurodegenerative diseases. Cells 9(9):2022. https://doi.org/10.3390/cells9092022

Wauters F, Cornelissen T, Imberechts D, Martin S, Koentjoro B, Sue C, Vangheluwe P, Vandenberghe W (2020) LRRK2 mutations impair depolarization-induced mitophagy through inhibition of mitochondrial accumulation of RAB10. Autophagy 16:203–222

William Langston J, Ballard P, Tetrud JW, Irwin I (1983) Chronic parkinsonism in humans due to a product of meperidine-analog synthesis. Science 219:979–980

Williams TI, Lynn BC, Markesbery WR, Lovell MA (2006) Increased levels of 4-hydroxynonenal and acrolein, neurotoxic markers of lipid peroxidation, in the brain in mild cognitive impairment and early Alzheimer's disease. Neurobiol Aging 27:1094–1099

Winslow AR, Chen CW, Corrochano S et al (2010) α-Synuclein impairs macroautophagy: implications for Parkinson's disease. J Cell Biol 190:1023–1037

Wirth M, Benson G, Schwarz C et al (2018) The effect of spermidine on memory performance in older adults at risk for dementia: a randomized controlled trial. Cortex 109:181–188

Xie Z, Klionsky DJ (2007) Autophagosome formation: core machinery and adaptations. Nat Cell Biol 9:1102–1109

Xilouri M, Vogiatzi T, Vekrellis K, Park D, Stefanis L (2009) Abberant α-synuclein confers toxicity to neurons in part through inhibition of chaperone-mediated autophagy. PLoS One 4:e5515

Xu CY, Kang WY, Chen YM, Jiang TF, Zhang J, Zhang LN, Ding JQ, Liu J, Di CS (2017) DJ-1 inhibits α-synuclein aggregation by regulating chaperone-mediated autophagy. Front Aging Neurosci 9:308. https://doi.org/10.3389/fnagi.2017.00308

Yamamoto S, Kuramoto K, Wang N, Situ X, Priyadarshini M, Zhang W, Cordoba-Chacon J, Layden BT, He C (2018) Autophagy differentially regulates insulin production and insulin sensitivity. Cell Rep 23:3286–3299

Yang Y, Zhang L (2020) The effects of caloric restriction and its mimetics in Alzheimer's disease through autophagy pathways. Food Funct 11:1211–1224

Yue Z, Yang XW (2013) Dangerous duet: LRRK2 and α-synuclein jam at CMA. Nat Neurosci 16: 375–377

Zhang L, Wang L, Wang R, Gao Y, Che H, Pan Y, Fu P (2017) Evaluating the effectiveness of GTM-1, rapamycin, and carbamazepine on autophagy and Alzheimer disease. Med Sci Monit 23:801–808

Zhao Y, Zhao B (2013) Oxidative stress and the pathogenesis of Alzheimer's disease. Oxidative Med Cell Longev 2013:316523. https://doi.org/10.1155/2013/316523

Zhu Y, Wang C, Yu M, Cui J, Liu L, Xu Z (2013) ULK1 and JNK are involved in mitophagy incurred by LRRK2 G2019S expression. Protein Cell 4:711–721

Chapter 14
Nanoparticles in Food Additives and Brain Health

Salma El-Shafie and Andreas Kakarougkas

Today, the use of nanotechnology is steeply expanding in almost every field of science owing to the unique chemical, physical, electrical, and mechanical properties of a material at the nanoscale compared to the corresponding bulk counterpart. The nanotechnology industry has grown from a 10-billion dollar enterprise in 2012 to over 1 trillion in 2015 (Li et al. 2020).

Nanoparticles, also known as ultra-fine particles, are defined as particles of matter between 1 and 100 nm in diameter, although sometimes the term is used for larger particles up to 500 nm, which is why in this review nanoparticles up to this size will be discussed for a more comprehensive evaluation. In the field of food science and food technology, nanoparticles are involved in several areas: starting from the cultivation of plants using nano-encapsulated fertilizers and pesticides, food processing by adding colour, texture, flavour and aroma; preventing spoilage, smart food packaging that can preserve shelf life and detect microorganisms, contaminants or toxic chemicals for food safety measures; and 'nutriceuticals' a term that combines nutrition with pharmaceuticals and is defined as food or part of food that gives medical and health benefits such as disease prevention or treatment.

The same unique physicochemical properties that make these new materials promising for many applications might also affect human health as they are not metabolized and are reported to accumulate in multiple organs after human exposure from several sources in the environment and via different routes, mainly inhalation, ingestion and dermal routes, emphasizing a need for evaluation of their health impacts at relevant exposure levels.

In this review, the focus will be on the foodborne nanoparticles that we directly consume via oral ingestion. Among the most common food additives that have been in use in the food industry for decades and all contain high percentages of

S. El-Shafie (✉) · A. Kakarougkas
Biology Department, School of Sciences and Engineering, The American University in Cairo, Cairo, Egypt
e-mail: salma4@aucegypt.edu

M. Salama (ed.), *Nutrigenomics and the Brain*, Nutritional Neurosciences,
https://doi.org/10.1007/978-981-16-9205-5_14

nanoparticles are titanium dioxide (E171), silver (E174) and silicon dioxide (E551). TiO_2 and SiO_2 are among the top produced nanomaterials by mass, and AgNPs make up to a quarter of the most popularly advertised nanomaterials (Ghebretatios et al. 2021; Vance et al. 2015).

Food additives are common in ultra-processed foods which are highly consumed in a 'western diet', which is a diet low in fibre and high in fat and carbohydrate. These ultra-processed foods include sweets, confectionaries, cookies, pastries, salad dressings, instant soups, noodles, fatty packaged snacks, powdered food and drinks and even baby formula. These additives can act as colorants, emulsifiers, sweeteners, thickeners, foaming or anti-foaming, and anti-caking agents giving enhanced sensory properties. In addition to this, they are highly profitable as some food additives can extend the shelf-life, protect the original appealing properties and prevent microbial proliferation in foods, which explains how these food additives and their nano-borne particles revolutionized the food industry. The percentage of nanoparticles in these common food additives is not fixed or standardized but differs from one supplier to another as will be described in the specific sections below. If the percentage of nanoparticles between 1 and 100 nm in the additive is 50% or more, the additive is called food-grade nanomaterial according to EU recommendations in 2011.

14.1 TiO_2

Titanium dioxide is one of the most popular pigments or colorants in use. Explaining its heavy use is its brightness and resistance to discoloration, and that is a desirable distinctive or non-common feature, white colour. It is estimated that approximately nine million tonnes of this pigment are produced annually worldwide, and that is 70% of the total produced pigments (Peters et al. 2018). The typical range of particle sizes in TiO_2 pigments is 50–500 nm. The white colour is best produced by particles within the range 200–300 nm; however, the Titanium Dioxide Manufacturers Association (TDMA) clearly states that the production may include particles <100 nm which is the size of a typical nanoparticle. Several groups have quantified the amount of 'true' nanoparticles in obtained TiO_2 pigments, and the percentages have ranged between 10 and 35% in different studies from different suppliers (Rompelberg et al. 2016). While it is heavily used in paints, coatings, plastics, inks, papers, cosmetics, sunscreens and toothpastes for its whiteness, this is also a desirable feature in several confectionaries, icings, dressings, sauces, chewing gum, candies and food products that have made TiO_2 a common food additive. It is also heavily used in the pharmaceutical industry as an opacity agent.

TiO_2 (E171) has been approved for use in food since 1966 allowing up to 1% of food weight. Other than the dynamic range in size of its constituent particles, it can also exist as anatase or rutile crystals or a mix of both. However, it has been accepted as a food additive in the anatase form first, and that was for several decades before its rutile form has been accepted as a food pigment in 2004, explaining why anatase is the most common in food products as was shown in a study that characterized samples of food additives from different suppliers to the USA and EU that found that the majority of samples were in the anatase form (Rompelberg et al. 2016).

It is estimated that the daily consumption of E171 additive may reach several hundreds of milligrams where a considerable fraction, around 36%, is nanoparticles. Different studies estimating the daily oral intake of TiO_2 nanoparticles, specifically, report variable numbers, where differences between oral intake among different age groups are observed. A Dutch study estimated the mean estimated long-term intake of TiO_2 NPs (extrapolated from the total TiO_2 amount in food products) ranged from 0.2 to 2 μg/kg bw/day, with the highest intake in young children where top products were swallowed toothpaste, sweets and candies, while for older groups, it was coffee creamers, sauces and dressings (Rompelberg et al. 2016). Worth noting is that the mass of ingested TiO_2 nanoparticles was calculated by calculating the percentage of mass of the fraction of NP <100 nm and multiplying the results of the TiO_2 intake estimations from food products by this number. In other words, the fraction of NP <100 nm was calculated as 0.31% by mass of the total TiO_2, so the mass of total TiO_2 NP uptake was calculated by multiplying the total TIO_2 intake by 0.0031. This is an important point as in several toxicology studies, it was found that health effects (such as inflammation, genotoxicity, and histological changes) were dependent on surface area (while size was constant) and not particle mass (Hoet et al. 2004). This is also emphasized in effects of nanoparticles via the inhalation route where tumour incidence correlated more strongly with NP surface area than with particle mass. Therefore, not only estimated ingested mass of consumed NPs is important to estimate but the total surface area as well.

A careful health risk assessment of TiO_2 NPs is lagging behind its explosive use. In the last decade, there has been rising concern of the toxicity of its long-term exposure based on evidence of possible adverse effects. The European Food Safety Authority (EFSA) panel, in a recent publication, concluded that the absorption and bioavailability of orally administered TiO_2 NPs are low, as most of the dose is eliminated in faeces, except for 0.1% which is absorbed by the gut and is distributed to different organs which was considered unlikely to cause a health hazard (EFSA Panel on Food Additives and Nutrient Sources added to Food (ANS) 2016a). However, accumulating evidence is implicating several adverse effects of these NPs on different body organs and systems driving diseases. Recent research findings on its impact on gut flora and induction of carcinogenesis in rats after oral administration of the TiO_2 additive which contains NPs triggered the ban of E171 food additive because of its possible harmful effects on humans in France, as the first country (Musial et al. 2020). The European Union commission is reported to be considering a complete ban or massive restrictions on titanium dioxide in food with a view on protecting children.

The effect of TiO_2 NP on the central nervous system has recently been examined in in vitro and in vivo studies.

during adulthood (Cui et al. 2014). All these are examples of how TIO_2 NPs can reach the brain through blood through the placenta.

In mice that were administered a single but high dose (5 g/kg weight) or 25 or 80 nm TIO_2 NP, brain lesions were found to be associated with exposure. The neurons in the brain hippocampus showed vacuoles that were greater in those administered 80 nm NPs and was attributed to fatty degeneration induced in the hippocampus of brain tissue (Wang et al. 2007).

A study on the tissue distribution and elimination after oral or IV administration of different TiO_2 NPs highlighted the slow tissue elimination. While the oral administration leads to low bioavailability, it is stressed that limited uptake in combination with slow elimination could result in potential tissue accumulation on the long run, particularly with daily exposure. Even with the IV administration that showed high tissue distribution in all tested organs that included the brain, measured titanium levels in faeces were indistinguishable compared to placebo-treated controls, and also no increase in titanium levels in urine was observed which supports the lack of elimination of the NPs (Geraets et al. 2014). Expected accumulation of NPs with our daily exposure, as a consequence of the negligible elimination, could pose a potential concern for human health risk.

Studies using the oral administration route in the literature are outnumbered by studies using inhalation routes as the original interest in TIO_2 is for its health effects due to its direct inhalation from air. However, these studies may be used to support the possible effects of TiO_2 when it reaches the blood stream, through any route and can accumulate systemically. A study that administered TiO_2 NPs through chronic intratracheal instillation showed lower percentage of brain to body weight and upon histological investigation, inflammatory cell aggregation and cell necrosis in the brain zones. It was shown with ICP-MS that after intratracheal instillation with TiO_2 NPs, these NPs were transported into the blood, and then they passed through the BBB and finally accumulated in the brain. In another study using aerosol inhalation of TIO_2 NPs, cerebral injury was induced 72 h post treatment in a dose-dependent effect in the case of the 10 or 20 nm NPs but not 200 nm.

A study on TiO_2 biodistribution after IV administration of (P25, 75% anatase and 25% rutile) in rats showed that there was biopersistence of the NPs in the liver, spleen and lungs up to 1 year treatment showing very low clearance rate, and interestingly, it was reported for the time that the presence of NPs in distal organs was linked to dysregulation of BBB physiology and neuroinflammation through serum mediators, which is a new way for TiO_2 NPs in exerting indirect effects on CNS (Disdier et al. 2015). Along the same lines, in a proteomic-based study on effects of subcutaneous injection of 25 nm TIO_2 NPs on mice, protein profiles in the exposed group's brains, particularly the lower activities of antioxidant enzymes and acetylcholine esterases alterations, seemed to be an indirect effect as the NPs were not detected in the brain. This indicates that TiO_2 NPs can affect brain health indirectly in addition to the posed hazards of its accumulation in the brain (Jeon et al. 2011). In another study that administered 5 nm TIO_2 NPs to mice by injection in abdominal cavity, NPs were detected in the brain and significantly decreased glutamate level in the brain. It is known that oxidative stress is related to excitotoxicity. Glutamate is an important neurotransmitter representing 30% of the total excitatory neurotransmitters in the brain. Furthermore, acetylcholinesterase activity was reduced. The implications of this could involve impacting the central cholinergic system which is responsible for regulation of cognitive functions (Ma et al. 2010). Exposure to TIO_2 NPs induced some neurons to turn into filamentous or inflammatory cells.

There is one in-vivo study that addressed the oral route of exposure to TIO2 NPs where TiO_2 NPs were orally administered to rats for 90 days at doses which were reported as safe showed accumulation of NPs in the brain, increased induced overexpression of cytokines in hippocampus and impairment of spatial memory of the rats evidenced by maze experiments. As this study directly locates TiO_2 in the brain after oral exposure and links it to neuroinflammation and behavioural changes in vivo, this drew attention to the use of TIO_2 NPs in the food industry and possible long-term effects on the CNS.

Furthermore, an additional way for possibly affecting brain health is through altering the gut microbiome which is a pillar in the gut-brain axis. TiO_2 NPs are reported to induce chronic changes in the composition or metabolic activity of gut microbiome. In a mouse model, oral administration of 100 mg/kg of TiO_2 NPs for 28 days resulted in increased proportion of the potentially harmful Actinobacteria and Proteobacteria and decreased proportion of beneficial Firmicutes and Bacteroidetes (Li et al. 2018). Another study using lower doses (2–50 mg/kg) more relevant to human exposure also supported these effects (Pinget et al. 2019). While another study also using low dose claiming to reflect relevant human exposure in food (2.5 mg/kg) didn't show faecal microbial diversity (Chen et al. 2017).

14.1.3 In Humans

In a human study, healthy volunteers ingested orally 23 or 46 mg of TIO_2 powder as either 160 or 380 nm nanoparticles. This resulted in blood levels of 50–100 μg/L with higher uptake observed for the smaller nanoparticle sample (Böckmann et al. 2000). While this appears as a small uptake, compared to the basal titanium level of 5–10 μg/L, this could be significant when ingested daily, and as reported extensively, it accumulates in organs and is not eliminated.

The significance of this study is that it is on humans and that it administered TIO_2 NPs orally which is the route of interest in this review as there are more studies and results on the different administration routes in animal studies mainly IV injection and intranasal installation, where the latter could be more relevant to assessing the risk of inhaling TIO_2 also present in the ambient air from traffic exhaust and tire brake release, but not relevant to the oral route of administration of assessing the risk of TIO_2 NP in food additives. Furthermore, the dose used is within the range of normal daily estimated intake for engineered nanoparticles. Recently, human intake of the food additive TiO_2 from food was estimated at 3 mg/kg bw/day (Dekkers et al. 2011), which for an average weight adult of 60 kg means daily intake of 180 mg, of which ~36% is NPs yielding 65 mg of TiO_2 NPs which is even higher than the dose used in this volunteer study.

Another human study showed that TIO_2 NPs were detected in blood of healthy volunteers starting at 2 h post administration of 100 mg of food-grade TiO_2 (which isn't all NPs like the earlier study) peaking at 6 h. Detection was confirmed by dark-field spectroscopy which identifies the presence of reflectant particles and inductively coupled plasma mass spectrometry ICP-MS which identifies total titanium (Pele et al. 2015). Again, the dose was relevant to the oral exposure in humans as it is within—here, as the maximum of—the estimated normal daily intake. As TiO_2 is known to exist in several common foods consumed during a typical diet (very high in dairy products; not due to addition of TiO_2 but as residual from cow feed (Rompelberg et al. 2016)), to eliminate confounding from different participant diets, the participants were asked to avoid consuming any dairy products from lunchtime the day preceding the study, to take their TiO_2 capsule at 9 am on the day of the study (TiO_2 capsule ingestion and blood measurements) and all had a common 'typical' diet throughout the day. The capsules used were anatase with *d*50 of 260 nm (260 nm is the median diameter of NPs).

Titanium dioxide spherical nanoparticles of 100–200 nm size in the anatase form are typically found in gut tissue as resistant to gastrointestinal degradation (Lomer et al. 2002). This has been strongly associated with Crohn's disease, where the pattern of increasing TIO_2 exposure from foods in the West since 1966, mirrors the increasing prevalence of Crohn's disease. Furthermore, a diet low in calcium and low in exogenous microparticles (mainly low food additives) alleviated the symptoms of ileal Crohn's disease and reduced the Crohn's disease activity index in a recent clinical trial. This emphasizes the inflammatory reaction to TiO_2 or other microparticles, while TiO_2 has been regarded as biologically inert; there are other examples such as the inflammatory reactions to a Ti metal hip prosthesis. While this doesn't fall under neurodegeneration or brain disease, it is a good example showing that possible abnormal immune responsiveness could be found in genetically diverse human subjects. Crohn's disease, like many diseases, is thought to arise from a combination of genetic predisposition and environmental factors. It is possible that the genetic composition of patients with Crohn's disease is making them more susceptible to hazardous effects of NPs. Building on this example, higher absorption of particles in gastrointestinal tract (GI) is observed in diabetics, where rat models (with induced diabetes) had 100-fold increase in absorption of polystyrene particles compared to control non-diabetic group (Buzea et al. 2007). Inflammation also is observed to allow the uptake/translocation of large particles up to 20 μm in diameter. The absorption of particles in the gut depends on the size with minimal uptake for large particles. An interesting study demonstrated this using polystyrene particles and showed that 6.6% NPs were uptaken when size was 50 nm; a slight decrease (uptake of 5.8%) was observed with 100 nm, a dramatic decrease (uptake of 0.8%) with 1 μm particles and 0% with 3 μm particles. The size affected the time required to cross the colonic mucus layer, which after that reach the enterocytes, a type of epithelial cells of the small and large intestines which helps in absorption of nutrients, where the NPs can enter the bloodstream (capillaries) and lymphatic system and then reach several organs. A point worth noting is that the zeta potential of all food additive TIO_2 NPs is negative, and charge is important for the kinetics of particles in GI, where positively charged particles have been shown to be trapped in the negatively charged mucus, while negatively charged nanoparticles diffused across the mucus layer and interacted with the epithelial cells. Also, subjects who have respiratory and circulatory diseases have higher capillary permeability, allowing fast translocation of metallic or nonmetallic nanoparticles into circulation. This shows how people of different genetic composition or different health/disease status could be at different risks for developing neurological health effects of TiO_2.

14.2 Silver Nanoparticles

Silver nanoparticles (AgNPs) are considered the most commercialized engineered nanomaterials (ENMs) accounting for more than 50% of global ENM consumer products (Temizel-Sekeryan and Hicks 2020). They are used in 15 industries including electronics, medical, textile, cosmetics, food packaging and coatings for its antimicrobial, antifouling, deodorizing, stain-resisting and electrical conducting properties.

As for the food industry, silver nanoparticles can be found in the food additive E174 in chocolates, cakes and confectionaries for decoration, but much more commonly encountered in food packaging for its preservative properties as an antimicrobial that can increase shelf life and quality of food. Silver has been used throughout history for its antimicrobial properties since the Ancient Egyptian times. Furthermore, colloidal silver, which contains ~25% nanoparticles, is used commonly as dietary supplements and as alternative medicine in disease and to strengthen immunity.

The EU and US food safety authorities are applying AgNPs in food packaging in a prudent way due to the inability to make conclusive statements on their toxicity, while Australia on the other hand bans the use of AgNP in packaging. As for Asia, in Malaysia and Indonesia, there are currently no safety regulations or guidelines, while Korea is planning to establish guidelines within a few year (Istiqola and Syafiuddin 2020).

Worth noting, currently, the EU and US food safety authorities regulate AgNPs use such that permissible limits for migration is set as 0.05 mg/kg or 0.05 mg/L of Ag +, without regulations in terms of nanoparticle migration. The use of nanoparticles in packaging can be divided into two groups: improved packaging and active packaging, where in the former, nanoparticles are mixed into the polymer matrix for enhancing the gas barrier properties, while in the latter, nanoparticles interact directly with food allowing better preservation of food. Silver nanoparticles are used for active packaging for its potent antimicrobial property.

Recently, it has been established that AgNPs used in food packaging and fresh food containers migrate or leach into food. A study on the migration of silver from different types of nanocomposites into different food simulants of different pH levels detected both Ag+ and Ag nanoparticles in food and that the migration was food and temperature dependent (Echegoyen and Nerín 2013). This emphasizes that attention should be given to heating food in such nano-enabled packaging. In their study the total Ag+ migrated was below the set limit of 0.05 mg/kg. However, as emphasized, there is a paucity in studies for food safety authorities to make relevant limits for regulations, and we are now living in a time where almost every food item purchased, ready to eat, ready to cook and ready to use, is packaged for extended shelf life and preservation. Silver NPs have been used for improving the quality and safety in the storage of fruits, vegetables, meat, bread and nuts (Bożena et al. 2015; Istiqola and Syafiuddin 2020). Some of the desirable properties associated with the use of AgNPs in packaging by comparing food packaged in composite polymers vs. in composite polymers that incorporate AgNps include decreasing water vapour and lower oxygen permeability that preserves the freshness of food and antimicrobial activity as AgNPs are considered the most significant antibacterial agent reported in literature against both positive and negative gram bacteria, enhancing mechanical properties such as tensis strength and elongation at break which would maintain the integrity of the packaging for a better seal protecting against stress factors during shipping and for improved freshness indicators of food such as maintaining acceptable lower pH in meat, higher vitamin C content in some fruits and higher nutritional status (total antioxidants, protein, phenols and flavonoids) where food wrapped in AgNP-containing packaging maintained the same nutritional status for a week, while there was significant decrease in those wrapped in packaging without AgNPs, while the former also protected against loss of moisture content (that would otherwise lead to stored vegetables that become dry) (Istiqola and Syafiuddin 2020). These properties revolutionized the food industry where the annual production of silver nanoparticles was estimated at 500 tonnes per year in 2009 and is estimated to reach 800 tonnes in 2025 as it has greater marketing value and its presence in products is more advertised compared to other NPs due to its known antimicrobial properties (Calderón-Jiménez et al. 2017).

While the documented migration of silver nanoparticles per food item as per some studies is below the set threshold for safety, their extremely slow elimination as emphasized in the in vivo section raises flags for the need for reassessing the risks of long-term exposure to silver nanoparticles in food.

14.2.1 In Vitro

As will be demonstrated next in the in vivo section, AgNPs ingested orally (which is the route of exposure relevant to the use of AgNPs in the food industry in packaging and dietary supplements) are biodistributed to several organs including the brain. Therefore, different types of brain cells and components of the BBB are studied in in vitro models evaluating the effects of exposure to these particles. Different types of cells in the brain play different roles; therefore, in vitro studies exposing TIO_2 to these different cell lines are discussed. Astrocytes function in regulating metal homeostasis, regulating ROS status and supplying nutrients to neurons, and microglial cells are macrophage-like cells that protect against microorganism invasion and mediate the neuroinflammatory response; nerve cells (neurons) connect with other neurons and with each other to transmit signal through synapses.

First, it was demonstrated in an in vitro BBB model using primary rat brain microvessel endothelial cells (rBMEC) isolated from adult Sprague-Dawley rats that AgNPs of sizes 25, 40 or 80 nm induced cytotoxicity and expression of pro-inflammatory cytokines such as IL-1b, IL-2 and TNF-alpha and increased permeability as evidenced by increased transport of fluorescein across the rBMEC monolayers (Trickler et al. 2010). Another BBB in vitro model of a co-culture of rat microvessel endothelial cells and astrocytes attributed transcytosis to being the route of NP uptake (Tang et al. 2010). The neurotoxicity of 15 nm AgNPs in neuronal-like PC12 cell line was linked to redox status perturbation evidenced by decreased glutathione peroxidase (gpx1) expression (Wang et al. 2009). A study using 3–5 nm AgNPs on glial (astrocyte ATL and microglial BV-2) and neuron N2a cell lines supported the induction of an immune response through upregulating chemokines CXCL13 and MARCO in all cell lines by exposure to AgNPs, oxidative stress through the decrease of glutathione synthase expression in astrocytes and deregulation of Alzheimer's disease (AD) genes: where APP, which is the amyloid-β eta (Aβ) precursor responsible for neurodegenerative disorders such as (AD), was upregulated in all cell lines; low-density lipid protein (LDLR) was downregulated in all; and neprilysin (NEP) was downregulated in N2a where they both function in the degradation of Aβ plaques (Huang et al. 2015). Another study observed structural impacts of exposure to 20 nm AgNPs on rat cortical neurons in the loss of cytoskeletal β-tubulin and filamentous actin, inhibited neurite outgrowth, mitochondrial dysfunction leading to cell death and reduced expression and clustering of clusters of the presynaptic vesicle protein synaptophysin and the postsynaptic receptor density protein PSD-95 which was observed in all doses even at the lowest dose of 1 μg/mL where the somatic and neuritic morphologies were still maintained. The doses used (1–50 μg/mL) were chosen to evaluate the effects of chronic exposure at low and medium doses on viability, cytoskeletal structure, key synaptic proteins and neurite length, where other studies had previously reported hazardous effects with primary neuronal cells, where doses as low as 10 μg/mL were reported to inhibit sodium and potassium currents, and calcium homeostasis was reported to be disrupted at 5 μg/mL, and dopamine levels reduced at 50 μg/mL (Xu et al. 2013). Another study on primary rat cerebellar granule cells (CGCs) showed that the excitotoxicity of AgNps <100 nm (25–75 μg/mL) is mediated through activation of glutamatergic NMDA receptor (*N*-methyl-D-aspartate) which leads to significantly increased intracellular calcium uptake, mitochondrial dysfunction and increased ROS compared to controls and effects that were abolished with the use of MK-801, an inhibitor of NMDA receptors (Ziemińska et al. 2014). It is now established that AgNPs can enter mammalian cells through diffusion (translocation), endocytosis or phagocytosis, where the particles or $Ag+$ ions can generate ROS upon entering the cytoplasm. AgNPs are observed to accumulate in the mitochondria inducing mitochondrial dysfunction which implies a reduction in mitochondrial membrane potential, creating ROS which leads to the denaturation or damage of proteins and nucleic acids inside the cell. Furthermore, due to the positive charge of AgNPs and the negative charge of cellular membranes, AgNPs also interact with cell membrane proteins to activate signalling pathways leading to cell death. Another study focused on the neurotoxicity through mouse embryonic stem cells where AgNPs disturbed neural progenitor cell-specific differentiation through deregulation of the expression of neural ectoderm marker genes even at concentrations as low as 0.1 μg/mL (Liu et al. 2015). Interestingly, a study that treated hippocampal neuronal cells (HT22 cell line) with both AgNPs and selenium showed that viability is restored with selenium as it suppresses ROS generation and caspase-3 activation (Ma et al. 2015). This is significant as the brain is high in lipid content and particularly susceptible to oxidative stress.

14.2.2 In Vivo

Unlike TiO2 NPs, in vivo studies using oral administration of AgNPs as the route of exposure to NPs, which is most relevant to reviewing the hazardous effects of ingesting these NPs from food products or nutriceuticals, are common. First, studies showing the biodistribution and clearance of AgNPs will be discussed followed by addressing the pathological and behavioural effects due to AgNPs exposure.

A study that administered AgNPs 10 or 25 nm in 100 or 500 mg/kg doses orally to rats over 28 days followed by a recovery period of 4 months allowing clearance of the particles showed that silver content decreased in all tissues (liver, kidney, spleen, ovaries, blood) except for the brain and testes which did not clear well implicating that the blood barrier for both organs could play a role in obstructing the clearance. It was interesting to find that the low dose of either size of AgNPs did not show any clearance, compared to the high dose which showed some clearance over the 4 months (Lee et al. 2013). This could be explained by the agglomeration of the nanoparticles which is more common to happen in high doses and can therefore be more easily cleared. This supported an earlier study with similar setting of oral administration of AgNPs (<20 non-coated or <15 nm PVP-coated NPs) in 90 mg/kg doses to rats over 28 days that allowed a recovery of only 2 months; in this study, it was emphasized that rats treated with AgNPs or $AgNO_3$ (a non-NP source of Ag+) were detected in rat tissue showing that the nanoparticles are formed in vivo from silver salts. Silver content was detected in all organs with highest values in liver and spleen, but after the clearance period of 2 months, silver was cleared from all organs except for the brain and testis. These two studies showed that different sizes (10–25 nm), different coating status and high or low doses resulted in accumulation of the NPs in the brain. Another study of similar setting orally administered 1–25 nm AgNPs at low 5 or 10 mg/kg doses for 28 days followed by 2-month clearance period, while with this low-dose AgNPs were detected in the brain (as well as the other organs tested in the former studies), all silver content from all organs was reported to be cleared within the 2 months, this could be due to the low dose which was originally detected at day 29 as low or negligible amount (Mata et al. 2018). In another study of oral administration of 34 nm AgNPs for 2 months at 100 μg per day to white mice, followed by clearance period for 1 month, clearance of silver from the brain was very low (only 6% cleared during the month), while more than 80% was cleared from blood and liver (Antsiferova et al. 2015). The finding that up to 1200 ng/g of silver in brain was detected raises questions whether it can influence cognitive functions (Antsiferova et al. 2015).

As for the pathological/histological changes due to AgNP exposure, in a rat study where <10 nm AgNPs were administered intragastrically at 1 or 10 mg/kg for 2 weeks, neuron shrinkage and astrocyte cytoplasmic and foot swelling were observed in brain tissue, and cytokine IL-4 was increased in blood (Xu et al. 2015). Intragastric administration of AgNPs is relevant to this review as AgNPs in food products would pass to the GIT, and also the integrity and stability of AgNPs in real human saliva have been proven (Li and Cummins 2020). Another study reported that low dose of 0.2 mg/kg of 10 nm AgNPs orally administered to rats for 2 weeks caused ROS that lead to neuron degeneration through toxicity to the myelin sheath where increased lipid peroxidation and decreased thiol groups of cysteine residues in the myelin fraction—which are known to be among the most susceptible redox sensitive targets—were observed (Dąbrowska-Bouta et al. 2019). The authors of the study chose an environmentally relevant dose where they reported that the predictable no-effect concentration for water contamination with AgNPs is 0.04–0.1 mg/L and that the studied dose 0.2 mg/kg is equivalent to 0.02 mg/L.

As for the behavioural changes due to AgNP exposure, a study showed that chronic oral administration of 2 mg/kg of 34 nm AgNPs in C57Bl/6 male mice for up to 180 days impaired their long-term memory as evidenced by open field, elevated plus maze, light-dark box and contextual fear conditioning tasks (Anna Antsiferova et al. 2018). Another study tested exposure of BALB/C mice to 0.1 mg/kg dose of 20–30 nm AgNPs through injection which resulted in the impairment of learning, social behaviour and motor function as evidenced by Morris water maze test (MWM), three-chamber social apparatus test and Rotarod performance test (Greish et al. 2019). Long-term as well as short-term memory impairment were also reported in a study that orally administered 20 nm AgNPs at 1 or 30 mg/kg to rats for 28 days at both doses, where higher contents of silver in the hippocampus were found compared to the lateral cortex, and none was detected in cerebellum or frontal cortex, supporting why noncognitive activity was similar to controls in this study (Węsierska et al. 2018). To compare the effects of exposure of brain tissue to AgNPs or Ag+ ions, a study compared oral administration of doses of 9 mg Ag/kg bw/day of 14 nm AgNPs or $AgNO_3$ in female Wistar rats which demonstrated that they cause similar neurotoxic effects of affecting neurotransmitter levels, inducing apoptosis via the same pathways hinting that the effects of AgNPs exposure could be mediated through the ions it releases (Hadrup et al. 2012). However, the results of this study show that while both forms of Ag affected dopamine levels which are implicated in reward-driven learning and its disturbances linked to Parkinson disease (PD), AgNP affected 5-hydroxytryptamine (5-HT) which is implicated in motor, cognitive and affective functions and in prediction of negative rewards and Ag+ affected noradrenaline which is implicated in modulating synaptic transmission and its disturbances linked to PD as well. Also, in the literature, there are examples of evidence for and against AgNP-specific effects, meaning different effects of treatment with AgNPs or Ag+, but such studies are not in CNS relevant cell lines (Hadrup et al. 2012).

Another in vivo study-related AgNPs to shortened adult lifespan and compromised tolerance to oxidative stress in *Drosophila* models where exposure to sublethal AgNPs activates Nrf-2-ARE pathway, which is conserved pathway functioning in the maintenance of cellular redox homeostasis that regulates the expression of several antioxidant genes such as superoxide dismutases sod1 and sod2, catalase, gclc which encodes glutamate-cysteine ligase regulatory subunit (GCLC) involved in glutathione (GSH) biosynthetic pathway, glutathione transferases gstD and gstE in response to ROS. AgNP-induced ROS-mediated stress responses such as apoptosis, DNA damage and autophagy (Mao et al. 2018).

While it can be argued that the doses used in some in vivo and in vitro studies were very high and accordingly demonstrated toxicity, it is also demonstrated in studies that high doses of AgNPs have lower absorption because they tend to agglomerate in the digestive system, while lower and relevant dietary concentrations are more easily absorbed. This has been reported in two studies where higher nephrotoxicity in rats was induced by oral administration of 125 mg/kg of 250 nm AgNPs for 28 days compared to the higher 300 mg/kg dose (Gherkhbolagh et al. 2018), and in a microbiome study where oral administration of 9 mg/kg of 10 nm AgNP for 90 days resulted in more severe dysbiosis in gut microbiota compared to higher doses 18 or 36 mg/kg (Williams et al. 2015).

14.2.3 In Humans

Colloidal silver is being increasingly marketed as a remedy for a range of health problems and used as popular dietary supplement and immunity booster without sufficient scientific evidence. The definition of colloidal silver is a solution of silver particles in the range of 1–1000 nm and is typically taken orally, depending on the product the size and fraction of nanoparticles vary. A recent study showed that rigorous regulation of the use of such products is needed, where 14 of colloidal silver products were analysed and significant differences were found in the claimed concentrations and no current labelling of nanomaterial is used; this study found that the size distribution in most products fell between 1 and 100 nm (De Leersnyder et al. 2020).

In a case study, a cancer patient who used colloidal silver as alternative medicine ingesting 1 ounce of the product daily for 4 months developed epilepsy myoclonic status epilepticus, coma and persistent vegetative state that were thought to be related to silver neurotoxicity, and his brain autopsy showed microscopic evidence of diffuse Alzheimer type 2 astrocytosis and microglial activation and elevated silver content in the cerebral grey matter (Mirsattari et al. 2004). It was argued that increased neuronal excitability caused by excess cellular uptake of calcium which has been found to be induced by AgNPs in vitro could have caused the epileptic state after exposure to colloidal silver in this patient. Myoclonic status epilepticus has been reported in some neurodegenerative diseases such as Creutzfeldt–Jakob disease. While it has never been reported to be caused by silver toxicity, silver toxicity has supportive evidence to cause neurodegeneration which could in turn cause myoclonic status epilepticus. Silver-containing anti-smoking pills were also reported to be associated with developing seizures.

In conclusion, from the data discussed, the explosive use of AgNPs in food products along with the persistence of AgNPs in the brain for long periods, its ability to alter the BBB permeability, its induced neurotoxicity and neuroinflammation supported in several studies in vivo and in vitro along with human case studies, and attention should be drawn to re-evaluate the health risks of using this nanomaterial in consumable food products.

14.3 Silicon Nanoparticles

While the share of nanomaterials in the food industry is expanding from offering enhanced food properties such as colour, texture, consistency, taste and sensation, to offering enhanced mechanical and anti-microbial properties in food packaging, for example, AgNPs, where they can also serve as indicators of food freshness and quality, and more recently to fortify foods with nutrients and vitamins or enhance their bioavailability such as nano-Fe supplementation, the use of nano silicon dioxide (SiO2) as a food additive, similar to TiO_2 and in contrast to these novel developments of nanotechnology in the food industry, has been on the market as a food additive for around 50 years. Synthetic amorphous silica, also known as SAS, is used as food additive E551 for its anti-caking 'anti-lumping' properties or preventing moisture and poor flow in powders, spices, creamers, confectionary sugar, powdered soups and dry food and for thickening and carrying flavours in other foods such as pastes and for clarifying beverages and controlling their foaming. It has long been approved by FDA in the USA allowing up to 2% by weight, and as of 2011, the maximum is allowed up to 1% by weight by the EU (Winkler et al. 2016). The anti-caking or enhanced flow ability is possible as SAS particles act as spacer, keeping host powder particles apart from each other and preventing their agglomeration, and it also absorbs liquid on the surface of host powders. Actually, silica particles exist in large amounts in nature; however, their synthetic forms developed for use in the food industry are only produced through either a pyrogenic process which generates silica nanoparticles of ~10 nm size mainly used for dry foods and powders or a wet process that produces silica gel or hydrous silica (Winkler et al. 2016). The proportion of SiO_2 nanoparticles in E551 was estimated at 33% in a recent study, and the human intake of total SiO_2 from food was estimated at 9.4 mg/kg bd/day where nanosilica was estimated at 1.8 mg/kg bd/day (Dekkers et al. 2011; Musial et al. 2020).

Production of SAS is estimated at 500,000 tons per year produced in Europe only, and the annual consumption of SiNPs worldwide was estimated at one million tons in 2015; it is used in different industries such as food and pharmaceuticals where it is ingested and also in cosmetics, tyre and paint industries (Li et al. 2020). To put TiO_2 and SiO_2 possible sources and amounts of ingestion in comparison, it was reported that the proportion of silicon nanoparticles ingested through food, pharmaceuticals and toothpastes were 14%, 44% and 42%, while the proportions of TiO_2 particles ingested in the three categories are 47%, 47% and 5% (Lomer et al. 2002). In 2013, SiNPs became one of the three most produced nanomaterials worldwide, where until 2012, it was reported that nearly 1.5 million tons of SiNPs had been placed in the global market. Needless to mention, not all this vast amount goes into the food industry, as SiNPs are involved in different industries and according to a consumer products inventory database in 2015, 100 nano-based products from a total of 846 contained SiNPs (Murugadoss et al. 2017).

14.3.1 In Vitro

In order to assess the neurotoxicity of SiO_2 NPs, in vitro studies reflecting intestinal absorption, the BBB permeability and finally cells of the nervous system are relevant. Regarding absorption, a study examining the absorption of different nano- and micro-sized SiO_2 particles in an inverted gut sac experiment showed that smaller particle sizes below 100 nm lead to higher intestinal absorption (Yoshida et al. 2014).

In vitro and in vivo research on many types of NPs has shown that nanoparticles do cross the intestinal epithelium by transport through the enterocytes and also showed that toxicity depends on many different properties such as particle size, composition, solubility, crystal structure, surface charge, surface modifications and shape. This explains the different in vitro results of different studies using different commercial and home-made NPs. One chronic in vitro study showed that SiO_2 was internalized by the intestinal epithelial cells as nanoparticles, and while it was not toxic to the epithelial cells, it raises a flag as this may subsequently enter the circulation and be distributed to other body parts (McCracken et al. 2013) that might be more sensitive to the NPs.

As evidenced by a number of studies, nanoparticles are found to be a promising brain-targeting strategy passing the BBB that presents a challenge for delivering drugs to lesions within the CNS. This has been employed for carrying MRI contrast agents, chemotherapeutic agents and photosensitizers. This is considered evidence supporting the possibility that SiO_2 NPs can enter the brain even without having enough studies examining and identifying such NPs in brain samples or in vitro studies presenting the BBB permeability. The BBB was addressed in some in vivo studies as will be discussed. A related in vitro study on BBB permeability used astrocytoma U87 cells that are considered good models for astrocytes that perform a variety of functions among which is controlling the BBB and showed that SiNPs induced cytotoxicity and affected mitochondrial function and survival signalling (Lai et al. 2010). Another study showed that astrocytes isolated and cultured from Wister rats are more vulnerable than isolated neurons to 10–20 nm SiNPs in vitro supported by morphological changes and alterations in the secondary structure of proteins (Limón-Pacheco et al. 2020).

In vitro studies on neuronal cell lines such as PC12 showed uptake of 25 nm SiNPs after exposure and a dose-response increase in induction of autophagy through increase of Beclin1 and LC-II and inhibition of mTOR signalling (Xie and Wu 2016), while testing indirect exposure of PC12 to the supernatant of microglia that has been directly exposed to 20 nm SiNPs in another study did not show similar cytotoxic effects on PC12 (Xue et al. 2012). Microglia have been shown to take up SiNPs in a time-dependent manner after exposure where low levels of the NPs can adversely affect microglial function; it increases ROS and RNS production, changes expression of proinflammatory genes, activates the microglia and induces release of proinflammatory mediators which not only may adversely affect microglial function but may cause cytotoxicity to surrounding neurons (Chen et al. 2018; Choi et al. 2010). In another study, SiNPs enhanced the secretion of cytokines by microglia where some of these cytokines affected dopamine synthesis through suppressing Th expression which caused cytotoxicity to PC12 cells, and therefore this study showed the adverse effects of SiNPs on neurons through microglia in vitro (Xue et al. 2012). Moreover, the same was also observed for TiO_2 in that study. Glia represent the largest cell population of the CNS that have a direct interaction with the blood-brain barrier via their end feet and importantly completely wrap the most abundant excitatory synapse in the CNS between axons of granule cells and dendritic tree of Purkinje cells; they play vital roles in metabolic trafficking, neurotransmitter cycling and protection of neurons against pathophysiological insults. An interesting study used chick Bergmann glia as an established cell line of glia and SiNPs (9–31 nm) as a characterized model of nanosized particulate matter (PM), which is becoming evident that its exposure is associated with development of behavioural deficits such as learning and memory disabilities specially in children and neurodegenerative disorders for the elderly vulnerable population as it triggers neurotoxic responses suggesting an important role of glutamate-mediated neurotransmission (Wang et al. 2017).

While in this study SiNPs did not show a significant effect on cell survival until high doses of 192 μg/mL, it was observed that the SiNPs agglomerate in the vicinity of the endoplasmic reticulum which could trigger the protein misfolding response. Moreover, at non-cytotoxic doses, a time-dependent differential effect on protein translation at initiation and elongation was observed where SiNP treatment resulted in the regulation of the initiation phase of protein synthesis through eIF2α phosphorylation, with an expected decrease in the elongation rate of protein synthesis (Rodríguez-Campuzano et al. 2020).

It was suggested by the authors that such response of glial cells could be mediating the deleterious effects of PM in the brain, as the PM, here demonstrated as SiNPs, activates pro-inflammatory response, oxidative stress and unfolded protein response ultimately leading to neuronal death which would lead to disproportionate

Glu release overactivating its receptors that regulate the glial physiological roles mentioned above, therefore in an attempt to efficiently remove Glu from the synaptic vicinity the transporters excitatory amino acid transporters (EAAT) of which EAAT-1, also known as Na+-dependent Glu/aspartate transporter (GLAST) profusely expressed on the membrane on glia cells are used where Glu triggers signalling pathways that regulate these gene expressions at the transcriptional and translational levels where the authors showed the glial response to SiNPs by modifying their proteome presumably in an efforts to overcome SiNPs neurotoxic effect.

Interestingly, several studies have noted signs of neurodegeneration in vitro and in vivo after exposure to SiNPs. One study demonstrated that a 24-h treatment in both human SK-N-SH and mouse neuro2a neuroblastoma cells with SiNPs of 15 nm induced pathological signs of Alzheimer's disease including altered expression of amyloid precursor protein and neprilysin, increased phosphorylation of tau protein at positions Ser262 and Ser396 and activation of glycogen syntheses kinase (GSK)-3β. Exposure of PC12 dopaminergic neurons to 15 nm SiNPs induced apoptosis, triggered oxidative stress and disturbed cell cycle, and this study had an in vivo component that supported the in vitro results and showed that the SiNPs entered the brain and deposited in the striatum where dopamine was depleted through downregulation of tyrosine hydroxylase; therefore, this study addressed SiNPs as a potential risk for neurodegenerative diseases and shed light on the need for assessment of SiNPs neurotoxicity as their applications in biomedical fields and therefore exposures are recently expanding (Wu et al. 2011). Another study by the same group showed upregulation of a-synuclein through inhibition of the ubiquitin-proteasome system which is a hallmark of Parkinson's disease (Xie and Wu 2016).

Another aspect relating to neuronal cells is shown in an in vitro study where SiNPs were not shown to impair neuron cell viability, induce neuroinflammation or alter neurite outgrowth, but neuronal differentiation markers showed that SiNPs caused a reduction in neuronal differentiation induction, where the markers MAP, MAP-2 and PI3 kinases of mature neurons were used to assess neuronal differentiation in SH-SY5Y using retinoic acid after a 24-h exposure to SiNPs (Ducray et al. 2017).

14.3.2 *In Vivo*

First, acute exposure studies showed no mortality or adverse effects after single oral exposure to SAS particles in rodents up to 5000 mg/kg. Sub-acute exposure studies (28-day oral exposure of doses 100–1000 mg/kg) in Wistar rats also showed no adverse effects on clinical signs, food consumption, animal weight, organ weights, histology or behaviour (Winkler et al. 2016). A longer 90-day sub-chronic exposure in Charles River rats with SAS dose up to 3500 mg/kg included in the feed did not affect food consumption, growth or survival (Winkler et al. 2016). Another 90-day study in Wister rats with doses up to 4000 mg/kg in diet supported the previous finding and showed no adverse effects on general condition, survival, clinical, pathological and histological examinations (Winkler et al. 2016). All these studies were described in OECD reports based on old studies conducted in the 1960s, 1970s and 1980s. Then in more recent studies, adverse effects on the liver started to be recognized converging on a potential systemic hazard of SAS. In a 10-week oral exposure study in BALB/c mice, nano vs. micro-sized silica particles were prepared and included in feed as 1% wt/wt, which amounts to 1500 mg/kg per day, and it was noticed that the NP group showed significantly higher serum level of alanine aminotransferase which is a biomarker of liver injury compared to controls and compared to the micro-sized SAS exposed group, and an appearance of fatty liver was also noted in the NP group (So et al. 2008). In another study on Sprague-Dawley rats, two kinds of SAS NPs were provided in feed at doses between 100 and 2500 mg/kg for 28 days or 84 days for the highest doses. In the 28-day exposure, no adverse effects were observed, while in the longer exposure, periportal liver fibrosis was observed and was higher the arm that was exposed to NPs of smaller surface area per weight (200 m^2/g vs. 380) (Van der Zande et al. 2014). Another study on Fischer rats providing SAS at 0, 1.25, 2.5 and 5% wt/wt in feed for 103 weeks showed significantly reduced liver weights in the highest two groups, but it was also noted the body weights were lower compared to the controls and lower exposures and cases of hyperplastic nodules in the liver and pheochromcytomas in the adrenal gland were observed in the SAS-exposed groups but not in the controls (Takizawa et al. 1988). The same study was conducted on B6C3F1 mice and showed overall body weight reduction in groups of mice exposed to the 2.5 and 5% wt/wt SAS in feed groups for 93 weeks compared to controls and lower exposures (Takizawa et al. 1988). A major criticism addressed to these long-term studies was that the feed used in the experiments was not examined for possible nutritional imbalances, possible test substance-related effects on food consumption was not addressed, and the test material (SAS) was not properly described; how it was produced, the particles were not characterized in terms of composition, impurities, physicochemical properties and size distribution. It can't be unnoticed however that in both studies, the liver toxicities were never observed in the controls. However, for these reasons, although there are several oral repeated long exposure in vivo experiments for SAS, the European Commission Scientific Committee on Consumer Safety recently pointed out that these long-term studies in rats and mice cannot be considered adequate for risk assessment.

Later, fluorescent dye-labelled silica particles were used to study in vivo distribution and excretion through injection of 50, 100 and 200 nm SiNPs that showed trapped particles detected by fluorescence in the liver up to 4 weeks after a single injection, and NPs of different sizes were eliminated at different rates but were generally mostly and quickly through urine then more slowly through faeces (Cho et al. 2009). Another study of oral exposure showed that the particles were absorbed and localized in the liver, while those administered through injection mainly localized in the liver and spleen and showed that the NPs were excreted through both urine and faeces regardless of route of exposure (Fu et al. 2013). Another study in Sprague-Dawley rats of repeated oral exposure for 5 days amounting to a dose of 100 mg/kg and detected increase in silicon content in liver that went back to background level within 14 days while the same dose through intravenous injection showed much higher content as expected in the liver that persisted for around 90 days. However, another repeated oral exposure study for 4 days at a lower dose of 1 mg/kg in Charles River mice was detected by fluorescence and ICP-MS SiNPs not their dissolved form mainly in liver tissues, and interestingly in several non-gastrointestinal organs in the mice such as kidney, lung, spleen and most relevantly and strikingly as this has not been shown in in vivo studies, the brain. The authors showed that the physicochemical characteristics and surface structure of the fluorescent SiNPs are similar to the commercial particles making them suitable for monitoring the biological fate of these particles to which we are frequently exposed to through different products (Zane et al. 2015). The use of fluorescence and ICP-MS SiNPs supported the possibility of the NPs reaching the brain, as SiNPs have been previously detected in the brain in in vivo SAS exposure studies, even in control groups with no statistical significance between control and exposed groups (Van der Zande et al. 2014), so the specification of fluorescence here was useful. In a study that used food-grade SiNPs given as a single dose of 500 mg/kg via oral route for studying its biokinetics in rats showed elevated SiO_2 in kidneys, liver, lung and spleen where it returned to background levels in lung within 2 days, but accumulation was more prolonged in other organs (Lee et al. 2017). It also showed that their absorption rate was rapid in the presence of glucose compared to protein which should be taken into account when predicting potential toxicities of food-borne NPs. A study that evaluated repeated exposure to mesoporous SiNPs showed that even for such potentially biodegradable NPs, even at the final time point studied of 4 weeks, NPs were still not completely cleared and persisted in the liver and spleen.

These numerous studies emphasize several points; first that oral exposure to SiNPs or SAS that contains SiNPs can be absorbed, distributed systematically and can cause liver injury which is of high relevance to neurotoxicity as the liver is the principal organ involved in detoxification. The definition of neurotoxicity extends to include secondary effects after kidney or liver damage where adverse effects on the nervous system arise due to the increased susceptibility to toxic substances ("Neurotoxicity. Identifying and Controlling Poisons in the Nervous System" 1993). With poor liver function, our nervous system is even more vulnerable to the toxins we are exposed to in our environment that could have been detoxified by a functional liver, whether these are toxins we are exposed to in our food, pesticides and chemicals in household products to count a few.

Regarding the property of agglomeration of NPs, it was thought that food-borne NPs would form agglomerates of size greater than 100 nm in food matrices, and therefore any toxicity reported from the in vitro studies due to small NP size would not be of concern; however, it has been experimentally demonstrated in an in vitro system mimicking human gastric digestion that while SiNPs form agglomerates under gastric acidic conditions, by switching conditions to mimic the subsequent intestinal digestion stage, the large agglomerates readily disintegrate into nano-sized particles and as high as ~80% of orally ingested SAS would display NP size in the intestinal lumen (Peters et al. 2012).

From another aspect, an in vivo study on mice that were orally administrated 27 nm SiNPs for 28 days at a dose calculated and explained by the authors to be matching reasonable intake from infant food investigated gut microbiota by 16S rRNA gene sequencing, level of inflammation and tissue integrity of gut, assessed classical indicators involved in gut-brain, gut-liver and gut-lung axis where neurobehavioral functions specifically were evaluated by open-field test and Morris water maze was carried out. The study demonstrated that SiNPs significantly caused spatial learning and memory impairments, caused damage to intestinal tissue integrity although they didn't induce intestinal inflammation, the microbial diversity in the gut was unexpectedly enhanced with increased abundances of Firmicutes and Patescibacteria and was thought to be associated with gut-brain-specific chemical substances such as Vipr1 and Sstr2, and no apparent changes in liver or lung tissues suggested the absence of gut-liver or gut-lung axis regulation (Diao et al. 2021).

After reviewing studies that show possible brain distribution of SiNPs after oral exposure, a study that addressed a different route of exposure to SiNPs, intranasal installation for 1 or 7 days, showed that the amount of SiNPs in the brain significantly increased and was detected in a descending order in the following brain regions, olfactory bulb, striatum, hippocampus, brain stem, cerebellum and frontal cortex. While the first region can be explained by the fact that the olfactory nerve pathway bypasses the BBB, the high content found in the striatum and hippocampus were of utmost interest as it questions whether SiNPs should be seriously assessed for risk of developing neurodegenerative disease. Study on SiNPs oral and dermal exposure in mice didn't cause any deposition of NPs in the brain regions, hippocampus or striatum, and it was argued that maybe the period of 90 days is not long enough to assess neurotoxicity in vivo (Shim et al. 2014), as it is reported that professional workers exposed to NPs for an extended period ranging from 5 to 13 months showed pulmonary fibrosis and NPs were detected in pulmonary epithelial and mesothelial cells. It was also showed in this study that it was a small percentage of SiNPs that was absorbed through the intestine, possibly explaining why there was no induced BBB damage as tested by Evans blue. However, there is conflicting results regarding BBB integrity after SiNP exposure in in vivo experiments.

In a study comparing effects of exposure to SiNPs through intraperitoneal injection for 7 days in control rats to streptozotocin-induced diabetic model of rats showed that there was exacerbation of regional BBB disruption, oedema formation, cell injuries and lower local cerebral blood flow (CBF) in the diabetic rats suggesting that diabetics are more vulnerable to NP-induced brain damage than healthy rats drawing attention to public health problem of diabetics exposure to NPs in the food, environment, especially with prevalent NPs such as SiNPs that are present in desert dust and in the living environment not limited to occupational exposures alone (Jose Vicente Lafuente et al. 2012).

An in vivo study on the nematode *Caenorhabditis elegans* examined the neurotoxicity of five commonly used NPs: SiO_2, CeO_2, CuO, Al_2O_3 and TiO_2 where comprehensive phenotyping was possible and RNA-seq after Ag and SiNPs (10–20 nm) treatment was carried out as they had the most significant effect on locomotion velocity even at the lowest dose of 10 μg/mL for SiNPs, which requires higher brain function and neurotoxicity phenotyped by the number of head thrashes per minute in a swim test in liquid media, and the RNA-seq analysis showed downregulation in several biological processes such as cell growth and locomotion and pathways such as neuroactive ligand receptor interaction, Wnt and MAPK signalling and SiNPs-downregulated genes involved in innate response (Viau et al. 2020).

In order to directly asses the effects of SiNPs and TiO2 NPs on the brain, a study examined the effects of direct exposure of the brain to TiO_2 (spherical 22 nm, 85% anatase, 15% rutile) and SiO_2 (5–35 nm) through intracerebroventricular administration using a stereotaxic approach. While not representing a natural route of exposure, it allows assessing neurotoxicity quickly, similar to the classical intraperitoneal administration, and would disseminate the NPs to the entire brain, and the doses tested for both NPs 5 and 10 μg represent 5–100% lower than those detected in the brain of mice exposed to TiO_2 through intragastric route for 60 consecutive days and therefore were chosen to avoid toxic effect. The study showed that both treatments significantly impaired motor performances of the C57BI6 mice measured on a rotarod at 20 rpm or accelerating from 4 to 40 rpm, even several weeks after initial acute exposure. Brain histopathology showed microglial activation that seemed to grow throughout the brain in a time-dependent manner suggesting induction of a long-lasting neuroinflammation. Together, the results of the study indicated that both NPs could impair the locomotor ability and that the impairment could be attributed to an inflammatory process that was maintained even 8 weeks after exposure which highlights how future research is needed to investigate if other neurotoxicological consequences such as neurotransmitter levels are altered underline the necessity of more in vivo studies to better characterize NPs exposure effects on the brain for long-term and low-dose treatment (Balvay et al. 2013). The approach used in the study is suitable to address the conflicting and limited in vivo results on nanoparticle exposure where neurotoxicity is poorly understood or documented, yet concerns about their long-term exposure possibly contributing to the appearance of neurodegenerative diseases such as Parkinson's or Alzheimer's diseases exist; therefore, to address the neurotoxic effects of brain exposure if any, direct injection was deemed appropriate for the question asked.

In zebrafish, which has been applied as a model for many neurodegenerative diseases owing to its genetic homology with most human genes including neurodegenerative genes, the potential neurotoxic effects of SiNPs were explored by behavioural phenotyping where adult zebrafish exposed to 20–25 nm SiNPs for 2 weeks at doses of 100, 300 and 1000 μg/mL showed observable effects on disturbed light/dark preferences, dampening exploratory behaviour and inhibiting memory capability tested via a battery of behavioural tests: light/dark preference test, novel tank diving test and T maze test (Li et al. 2020). The results showed pronounced behavioural symptoms similar to neurodegenerative diseases, represented by elevated depression and anxiety in addition to inhibited learning and memory; therefore, the transcriptional alteration of autophagy and parkinsonism-related genes were assessed and showed a hormesis effect of SiNPs on regulation of autophagy genes, which describes the biphasic dose response with the low-dose stimulation showing beneficial effects, here upregulation of autophagy genes, and a high-dose stimulation showing inhibitory or toxic effects, here downregulation of autophagy genes and impairment of autophagy, which is considered a hallmark of Parkinson's disease.

Another study in zebrafish demonstrated the neurotoxic effect of different sizes, 15 and 50 nm and different concentrations; 300 and 1000 μg/mL of SiNPs on cognitive behaviour specifically learning and memory represented by colour discrimination and locomotive activity represented by distance, velocity, freezing time ratio; and 3D spatiotemporal reconstruction of the swim path (Li et al. 2014). As Parkinson's disease model in zebrafish exists, it is noted that decreased locomotive activity is the primary representative behavioural alteration exhibited in this model. Moreover, the authors showed that dopaminergic neurons (DA) identified by labelling with TH antibody were shown to be reduced in posterior tuberculum (PT) which is the counterpart of the substantia nigra in amniotes. Therefore the study suggests that SiNP treatment could induce characteristics of Parkinson's disease in zebrafish.

A study on zebrafish embryos microinjected with varying doses of 100 nm SiNPs showed that the NPs disrupt the axonal integrity, decrease lengths of axons and cytotoxicity in the brain, and CNS cells did not reach statistical significance, but microarray analysis of gene expression was carried out and revealed that SiNPs markedly downregulated genes associated with neural function and therefore affect signalling of the neuroactive ligand-receptor interaction pathway; this highlights the developmental neurotoxicity of SiNPs (Wei et al. 2020).

In conclusion, the available in vivo data are inconclusive for SiNPs where some studies show no significant absorption; others show significant absorption in several organs and vary in the clearance times, while others show toxic effects mainly on the liver which is of relevance in considering neurotoxicity. Worth noting is that old reports that relied on the finding that SiNPs like several NPs agglomerate to bodies of >100 nm size, and therefore the concerns over the potential toxicities of NPs should be brushed away are clarified where ~80% of ingested SiNPs that agglomerate are resolved to NPs, and similarly old reports on the lack of absorption in any organs are being challenged as the detection of the NPs is now possible through different techniques.

14.3.3 In Humans

There is limited knowledge on the tissue concentration of SiNPs in humans. One study that examined tissue concentrations of SiO_2 and TiO_2 NPs in humans post-mortem found that Si concentration in liver is 8 ± 8 mg Si/kg, which is 50% of the estimated Si liver concentration using a kinetic model developed by Van Kesteren et al. 21–23 mg Si/kg, based on consumer intake of SiO_2 from food data (9.4 mg Si/kg bw/day (Peters et al. 2020; Van Kesteren et al. 2015). Particulate SiO_2 represented about 10% of the total Si concentration in the human post-mortem samples. While silica is known to occur normally in the environment and therefore small amounts of SiNPs are normally expected in our bodies, in addition inhalation of silica dust is common both in working and living environments; however, the contribution is limited and so is dermal exposure to SiNPs; therefore, the authors conclude the plausibility that the source of SiO_2 and TiO_2 in the study is due to food, toothpaste and pharmaceuticals, all through oral exposure. While SiO_2 and TiO_2 are authorized ingredients in food and pharmaceuticals, knowledge about their resulting tissue concentrations in humans is extremely limited and that regarding particle size is virtually absent which is required for risk assessment.

Worth mentioning the estimated consumption by a European population of 9.4 mg per kg body weight and day is greater than the lowest observed adverse effect level (LOAEL) of 5 mg estimated based on the chronic oral exposure studies in mice where the no observed adverse effect (NOAEL) was 500 mg and the LOAEL (for liver atrophy) was around 8100–1500 mg/kg, by dividing the NOAEL by a default uncertainty factor of 100 for interspecies and interindividual differences in sensitivity, the NOAEL was set to 5 mg/kg in humans, which emphasizes that our dietary exposure to SAS should even be reduced to be considered safe (Winkler et al. 2016).

14.4 Regulations on the Use of Food Additives

The increasing use of TiO_2 (E171), AgNPs (E174) and SiNPs (E551) as food additives in various consumer products, their fate, and their safety and effects on human health are of growing concern. Manufacturers do not require mention of the NP characterization in products. It is clear from the in vitro and in vivo data reviewed in this chapter that different size and particle surface or physicochemical properties used in different studies lead to varying toxicological responses, and only limited studies use these food additive chemicals as they are not the isolated commercial NPs of specific size and characteristics which is not necessarily the same as what is present in the food additives we consume and are interested in evaluating their health effects specifically. This is why recently the European commission began to re-evaluate all the approved food additives and planned to complete the evaluation before 2020, during which particle size and relevant physicochemical characters of food additives and nanotechnology-based products (Commission Regulation (EU) 2010) are considered.

There is a growing amount of supporting evidence for neurobehavioral toxicity in animals emphasizing the need for considering the potential neurotoxicity of food additives that have been in use for decades.

It is the task of regulatory agencies to limit public exposure to toxic chemicals through programs mandated by law. Because of the great diversity of toxic substances, many statutes or acts exist to control their use. These acts are administered by various federal agencies, but primarily by the Environmental Protection Agency (EPA), the Food and Drug Administration (FDA) and the Occupational Safety and

Health Administration (OSHA). The act mostly related to food additives is the Federal Food, Drug, and Cosmetic Act (FFDCA) where the agency primarily responsible for it is FDA, and it regulates food, food additives, drugs and cosmetics. Neurotoxicity is generally not explicitly mentioned in legislation mandating the regulation of toxic substances; it is only included implicitly as a toxicity concern. Federal laws regulating toxic chemicals can be classified into three levels: first, licensing and registration laws which include requirements for toxicity testing; second, standard-setting laws which include recommended or required limits on toxic substances and dictating appropriate labelling of products that contain toxic chemicals; and third control-oriented measures which include chemical groups explicitly identified as targets of concern.

Unlike acts dealing with pesticides, FDA deals with food additives in less formal procedures specially in reviewing existing food additives. No monitoring or tracking of adverse effect or physician reports is done for food additives; few exceptions exist, such as the case of Aspartame, but most food additives are not a subject of formal reporting programs by FDA. However, while it is not required, FDA can track such information and use it to reassess the risks.

Moreover, FDA does not routinely require specific neurotoxicity testing for food additives as it is thought as unnecessary and that the general toxicity testing can adequately detect adverse effects on the nervous system, while other scientists think that these general toxicity profiles are not sensitive enough to detect many neurotoxic effects and that specific neurotoxicity testing is essential for a complete toxicity evaluation. Neurotoxicity refers to any toxic agent that adversely affects any of the structural or functional—including behavioural—components of the nervous system. It is evident that many of these adverse effects could be missed in general toxicity assays. Most substances however must be subjected to subchronic studies in mammals as well as reproductive tests, so some prediction of adult or developmental neurotoxicity, or both, may be provided or may be missed. The toxicity test requirements in a petition are negotiable, but the petitioner must present sufficient data to ensure a reasonable certainty that no harm will result from the use of the additive ("Neurotoxicity. Identifying and Controlling Poisons in the Nervous System" 1993).

There are four main principles published as guidelines for the approach of safety assessment of food additives by the FDA in 1982 in the publication 'Toxicological Principles for the Safety Assessment of Direct Food Additives and Color Additives Used in Food' which is commonly known as 'the Redbook'. Although there have been revisions to the Redbook, the overall approach for food safety assessment is still the same. First, some general toxicological information is required for each food additive; second, the level of concern (LOC) data is what dictates the amount of safety data required for this additive; third, LOC is based on the magnitude of human intake of the additive as well as its molecular structure; and fourth the evaluation of the additive may be adjusted if any adverse effects are found. Finally, results from toxicology studies are used to calculate the acceptable daily intake for this additive (ADI) which is compared to the estimated daily intake (EDI). If the EDI is less than the ADI, the additive is determined to be safe under the proposed conditions of use.

The ADI is calculated by specifying the most sensitive indicator of toxicity (which has to be a noncancer effect) in toxicological in vivo studies, then the highest NOAEL is identified for the effect and then corrected by the uncertainty/interspecies factor. For calculating the EDI, the dietary concentration of the food additive is multiplied by the total weight of the food consumed by the individual per day.

For titanium dioxide, silver and silicon dioxide food additives no ADI values are specified as there are limitations in the available data. The FDA decides however how much of the food additives to be added to food on the basis of the safety data submitted with the application to FDA where its interpretation of safety is drawn from the legislative history of the act 'The Food Additives Amendment of 1954' which used the phrase 'reasonable certainty of no harm' and has incorporated that standard into its regulations regarding toxicity testing. This is how the quantity of titanium dioxide is set not to exceed 1% by weight of the food and silica is used in an amount not to exceed 2% by weight of the food. Limiting the amount of NPs per food product is not enough to control the amount of NPs ingested due to different diets, food style and amounts of food consumed by individuals. After reviewing studies that show the absorption, distribution and adverse neurotoxicological in both short- and long-term studies in mammals, the neurotoxicity of these common food additives is realized as a public health concern. While the use of NPs is recently increasing tremendously in nanomedicine for desirable delivery to the brain due to their ability to cross the BBB, considering the susceptibility of the brain towards different kinds of injuries, due to its limited regenerative ability, NPs presence in the brain should be considered cautiously, as this could contribute to neurodegeneration. This highlights the need for giving attention to the neurotoxicity of NPs in vivo and assessing the effects of long-term small-dose exposure to NPs representing oral intake of it in food over lifetime, especially in the case of biopersistent metal nanoparticles such as TiO_2 and silver, and not as much the biocompatible SiO_2.

Clear specifications by FDA such as characterization of particle size to be used as food additives are required for these food additives and more in depth evaluation of the health effect of such NPs is needed to assess their safety. More research is needed before an ADI can be established which will therefore ensure the safety of consuming them in food by the public. Many scientists and public health observers believe that food additives should come under the same scrutiny as drugs, particularly because many of them are regularly ingested by millions of people.

An important aspect to consider in assessing the use and safety of NPs in food additives is the economic importance that arises in the regulation of toxic substances as the impact of the direct regulation can be costly on market prices for consumers and importantly on industry profits. Here, big industries are involved as the food and beverage sector is a global multi-trillion dollar industry, and a recent estimate of the global economic impact of nanotechnology was projected to be at least \$3 trillion by 2020, which may employ up to 6 million as labour in the rising nanotechnology industries worldwide (He et al. 2019). The value of nanotechnology in the food market, specifically, worldwide was estimated at \$20.4 billion by 2010 and \$1 trillion by 2013 in a report in 2006 (Food Safety Authority of Ireland 2006).

Therefore, some activist groups question the judgement of food scientists hired by the industry to make safety determinations or decisions as they could have an inherent conflict of interest. An industry group such as Flavor and Extract Manufacturers Association (FEMA) has a GRAS program where GRAS Generally Regarded as Safe is an acronym used by FDA to designate chemicals as safe according to experts. More than 100 countries accept FEMA's GRAS determinations which demonstrate a global confidence in its food regulatory program.

Under the European legislation in 2008, the safety of all food additives already authorized for use in the EU were planned to be re-evaluated with a deadline of completion of this process in 2020; however, given the large amount of work and research, the program is still ongoing for some additives. In 2021, the EFSA re-evaluation of TiO_2 highlighted potential immunotoxicity and inflammation with E171, potential neurotoxicity of TiO_2 NPs and induction of aberrant crypt foci in the colon which represent adverse effects. Regarding genotoxicity, recent research revised in the re-evaluation showed TIO_2 can induce DNA strand breaks and DNA damage; no size cutoff value regarding genotoxicity was identified, raising concern for genotoxicity in using E171, and given the many uncertainties, the Panel concluded that E 171 can no longer be considered as safe as a food additive (Younes et al. 2021).

For the re-evaluation of silver food additive (E174), the latest opinion publication by EFSA in 2017 noted there are several gaps and concerns needed to conduct a risk assessment as there is a lack of toxicity studies using the food additive; the particle size distribution in E 174 is not always given and the evidence of release of silver ions from the food additive is of concern, and the decision was that the information available based on reviewed studies was insufficient to assess its safety. The panel recommended that the specifications for E 174 should include the mean particle size and size distribution, as well as the percentage (in number not mass) of particles in the nanoscale (<100 nm), present in the food additive (EFSA Panel on Food Additives and Nutrient Sources added to Food (ANS) 2016b).

As for silicon, in the latest re-evaluation of E551 by EFSA, although there was no indication of adverse effects in the subchronic, reproductive and developmental toxicological studies, despite the addressed limitations in these studies, and without inducing genotoxicity, still EFSA did not give E 551 the safety clear as in the absence of a long-term study with nano silicon dioxide, the panel could not extrapolate the results from the available chronic study with a material, which does not cover the full-size range of the nanoparticles that could be present in the food additive E 551, to a material complying with the current specifications for E 551. The panel required clear characterization of particle size for E551. Because of the limitations in the available data, the current ADI is 'not specified' (Younes et al. 2018).

In conclusion, with the growing use of nanotechnology in the food industry and its projected growth in the coming years, exposure to NPs is unavoidable. By reviewing the available in vitro, in vivo and human research on the distribution and potential neurotoxicity of the three common food additives, and reviewing the raised flags in EFSA re-evaluations, it is of high importance to have more in vivo research that mimics the realistic exposure to these NPs through ingestion, in short- and long-term studies and assess different neurological toxicities as there are several limitations and criticisms in the existing in vivo data upon which ADI should be calculated. Another point is epidemiological studies investigating the associations between foodborne NPs and neurological damage and behavioural impairments is required, as most epidemiological data is based on the association between NPs in PM in air and not focusing on other prevalent sources of NPs such as food. More attention should be given to studying chemicals of potential neurotoxicities, as evidenced by research studies reviewed, that have wide exposure (all the public consumes food that could be containing food additives), it is a long exposure as the public would consume these products containing NPs over their lifetime, and the exposure is at low doses. This is similar to the definition of neurotoxins as chemical time bombs, characterized by long exposure at low doses, which leads to long latency neurotoxicity (Cory-Slechta 1989). This is worth addressing in further research, as several in vitro and in vivo studies showed that all three NPs lead to neurodegeneration.

References

Antsiferova A, Buzulukov Y, Demin V, Kashkarov P, Kovalchuk M, Petritskaya E (2015) Extremely low level of Ag nanoparticle excretion from mice brain in in vivo experiments. IOP Conf Ser Mater Sci Eng 98(1):012003. https://doi.org/10.1088/1757-899X/98/1/012003

Antsiferova A, Kopaeva M, Kashkarov P (2018) Effects of prolonged silver nanoparticle exposure on the contextual cognition and behavior of mammals. Materials 11(4):1–15. https://doi.org/10.3390/ma11040558

Balvay A, Thieriet N, Lakhdar L, Bencsik A (2013) Comparative study of neurologic effects of nano-TiO2 versus SiO2 after direct intracerebral exposure in mice. J Phys Conf Ser 429(1): 012027. https://doi.org/10.1088/1742-6596/429/1/012027

Böckmann J, Lahl H, Eckert T, Unterhalt B (2000) Titan-Blutspiegel vor und nach Belastungsversuchen mit Titandioxid [Blood titanium levels before and after oral administration titanium dioxide]. Pharmazie 55(2):140–143. http://www.ncbi.nlm.nih.gov/entrez/query.fcgi?cmd=Retrieve&db=PubMed&dopt=Citation&list_uids=10723775

Bożena WR, Adam E, Mieczyslaw O, Elżbieta B, Ewa G (2015) Application of silver nanoparticles to fruits and vegetables to improve their post-harvest shelf life. In: International multidisciplinary scientific geoconference surveying geology and mining ecology management, SGEM, vol 1(6), pp 283–290. https://doi.org/10.5593/sgem2015/b61/s25.039

Buzea C, Pacheco II, Robbie K (2007) Nanomaterials and nanoparticles: sources and toxicity. Biointerphases 2(4):MR17–MR71. https://doi.org/10.1116/1.2815690

Calderón-Jiménez B, Johnson ME, Montoro Bustos AR, Murphy KE, Winchester MR, Vega Baudrit JR (2017) Silver nanoparticles: technological advances societal impacts and metrological challenges. Front Chem 5:6. https://doi.org/10.3389/fchem.2017.00006

Chen H, Zhao R, Wang B, Cai C, Zheng L, Wang H, Wang M, Ouyang H, Zhou X, Chai Z, Zhao Y, Feng W (2017) The effects of orally administered Ag, TiO2 and SiO2 nanoparticles on gut microbiota composition and colitis induction in mice. NanoImpact 8:80–88. https://doi.org/10.1016/j.impact.2017.07.005

Chen L, Liu J, Zhang Y, Zhang G, Kang Y, Chen A, Feng X, Shao L (2018) The toxicity of silica nanoparticles to the immune system. Nanomedicine 13(15):1939–1962. https://doi.org/10.2217/nnm-2018-0076

Cho M, Cho WS, Choi M, Kim SJ, Han BS, Kim SH, Kim HO, Sheen YY, Jeong J (2009) The impact of size on tissue distribution and elimination by single intravenous injection of silica nanoparticles. Toxicol Lett 189(3):177–183. https://doi.org/10.1016/j.toxlet.2009.04.017

Choi J, Zheng Q, Katz HE, Guilarte TR (2010) Silica-based nanoparticle uptake and cellular response by primary microglia. Environ Health Perspect 118(5):589–595. https://doi.org/10.1289/ehp.0901534

Cory-Slechta DA (1989) Behavioral measures of neurotoxicity. Neurotoxicology 10(2):271–296. https://doi.org/10.17226/1352

Cui Y, Chen X, Zhou Z, Lei Y, Ma M, Cao R, Sun T, Xu J, Huo M, Cao R, Wen C, Che Y (2014) Prenatal exposure to nanoparticulate titanium dioxide enhances depressive-like behaviors in adult rats. Chemosphere 96:99–104. https://doi.org/10.1016/j.chemosphere.2013.07.051

Dąbrowska-Bouta B, Sulkowski G, Strużyński W, Strużyńska L (2019) Prolonged exposure to silver nanoparticles results in oxidative stress in cerebral myelin. Neurotox Res 35(3):495–504. https://doi.org/10.1007/s12640-018-9977-0

De Leersnyder I, Rijckaert H, De Gelder L, Van Driessche I, Vermeir P (2020) High variability in silver particle characteristics, silver concentrations, and production batches of commercially available products indicates the need for a more rigorous approach. Nanomaterials 10(7):1–22. https://doi.org/10.3390/nano10071394

Dekkers S, Krystek P, Peters RJB, Lankveld DPK, Bokkers BGH, Van Hoeven-Arentzen PH, Bouwmeester H, Oomen AG (2011) Presence and risks of nanosilica in food products. Nanotoxicology 5(3):393–405. https://doi.org/10.3109/17435390.2010.519836

Diao J, Xia Y, Jiang X, Qiu J, Cheng S, Su J, Duan X, Gao M, Qin X, Zhang J, Fan J, Zou Z, Chen C (2021) Silicon dioxide nanoparticles induced neurobehavioral impairments by disrupting microbiota–gut–brain axis. J Nanobiotechnol 19(1):1–20. https://doi.org/10.1186/s12951-021-00916-2

Disdier C, Devoy J, Cosnefroy A, Chalansonnet M, Herlin-Boime N, Brun E, Lund A, Mabondzo A (2015) Tissue biodistribution of intravenously administrated titanium dioxide nanoparticles revealed blood-brain barrier clearance and brain inflammation in rat. Part Fibre Toxicol 12(1):1–20. https://doi.org/10.1186/s12989-015-0102-8

Ducray AD, Stojiljkovic A, Möller A, Stoffel MH, Widmer HR, Frenz M, Mevissen M (2017) Uptake of silica nanoparticles in the brain and effects on neuronal differentiation using different in vitro models. Nanomedicine 13(3):1195–1204. https://doi.org/10.1016/j.nano.2016.11.001

Echegoyen Y, Nerín C (2013) Nanoparticle release from nano-silver antimicrobial food containers. Food Chem Toxicol 62:16–22. https://doi.org/10.1016/j.fct.2013.08.014

EFSA Panel on Food Additives and Nutrient Sources added to Food (ANS) (2016a) Re-evaluation of titanium dioxide (E 171) as a food additive. EFSA J 14(9):e04545. https://doi.org/10.2903/j.efsa.2016.4545

EFSA Panel on Food Additives and Nutrient Sources added to Food (ANS) (2016b) Scientific opinion on the re-evaluation of silver (E 174) as food additive. EFSA J 14(1):4364. https://doi.org/10.2903/j.efsa.2016.4364

Food Safety Authority of Ireland (2006) The relevance for food safety of applications of nanotechnology. file:///C:/Users/giuliana.vozza.ICTDOMAIN/Downloads/Nanotechnology_report(2).pdf

Fu C, Liu T, Li L, Liu H, Chen D, Tang F (2013) The absorption, distribution, excretion and toxicity of mesoporous silica nanoparticles in mice following different exposure routes. Biomaterials 34(10):2565–2575. https://doi.org/10.1016/j.biomaterials.2012.12.043

Geraets L, Oomen AG, Krystek P, Jacobsen NR, Wallin H, Laurentie M, Verharen HW, Brandon EFA, de Jong WH (2014) Tissue distribution and elimination after oral and intravenous administration of different titanium dioxide nanoparticles in rats. Part Fibre Toxicol 11(1):1–21. https://doi.org/10.1186/1743-8977-11-30

Ghebretatios M, Schaly S, Prakash S (2021) Nanoparticles in the food industry and their impact on human gut microbiome and diseases. Int J Mol Sci 22(4):1–24. https://doi.org/10.3390/ijms22041942

Gherkhbolagh MH, Alizadeh Z, Asari MJ, Sohrabi M (2018) In vivo induced nephrotoxicity of silver nanoparticles in rat after oral administration. J Res Med Dent Sci 6(1):43–51. https://doi.org/10.24896/jrmds.2018619

Greish K, Alqahtani AA, Alotaibi AF, Abdulla AM, Bukelly AT, Alsobyani FM, Alharbi GH, Alkiyumi IS, Aldawish MM, Alshahrani TF, Pittalà V, Taurin S, Kamal A (2019) The effect of silver nanoparticles on learning, memory and social interaction in balb/c mice. Int J Environ Res Public Health 16(1):148. https://doi.org/10.3390/ijerph16010148

Hadrup N, Loeschner K, Mortensen A, Sharma AK, Qvortrup K, Larsen EH, Lam HR (2012) The similar neurotoxic effects of nanoparticulate and ionic silver in vivo and in vitro. Neurotoxicology 33(3):416–423. https://doi.org/10.1016/j.neuro.2012.04.008

He X, Deng H, Hwang H, min. (2019) The current application of nanotechnology in food and agriculture. J Food Drug Anal 27(1):1–21. https://doi.org/10.1016/j.jfda.2018.12.002

Hoet PH, Brüske-Hohlfeld I, Salata OV (2004) Nanoparticles—known and unknown health risks. J Nanobiotechnol 2(1):12. https://doi.org/10.1186/1477-3155-2-12

Hu R, Zheng L, Zhang T, Gao G, Cui Y, Cheng Z, Cheng J, Hong M, Tang M, Hong F (2011) Molecular mechanism of hippocampal apoptosis of mice following exposure to titanium dioxide nanoparticles. J Hazard Mater 191(1–3):32–40. https://doi.org/10.1016/j.jhazmat.2011.04.027

Huang CL, Hsiao IL, Lin HC, Wang CF, Huang YJ, Chuang CY (2015) Silver nanoparticles affect on gene expression of inflammatory and neurodegenerative responses in mouse brain neural cells. Environ Res 136:253–263. https://doi.org/10.1016/j.envres.2014.11.006

Istiqola A, Syafiuddin A (2020) A review of silver nanoparticles in food packaging technologies: regulation, methods, properties, migration, and future challenges. J Chin Chem Soc 67(11): 1942–1956. https://doi.org/10.1002/jccs.202000179

Jeon YM, Park SK, Lee MY (2011) Toxicoproteomic identification of TiO2 nanoparticle-induced protein expression changes in mouse brain. Anim Cells Syst 15(2):107–114. https://doi.org/10.1080/19768354.2011.555144

Koeneman BA, Zhang Y, Westerhoff P, Chen Y, Crittenden JC, Capco DG (2010) Toxicity and cellular responses of intestinal cells exposed to titanium dioxide. Cell Biol Toxicol 26(3): 225–238. https://doi.org/10.1007/s10565-009-9132-z

Lai JCK, Ananthakrishnan G, Jandhyam S, Dukhande VV, Bhushan A, Gokhale M, Daniels CK, Leung SW (2010) Treatment of human astrocytoma U87 cells with silicon dioxide nanoparticles lowers their survival and alters their expression of mitochondrial and cell signaling proteins. Int J Nanomedicine 5:715–723. https://doi.org/10.2147/IJN.S5238

Lee JH, Kim YS, Song KS, Ryu HR, Sung JH, Park JD, Park HM, Song NW, Shin BS, Marshak D, Ahn K, Lee JE, Yu IJ (2013) Biopersistence of silver nanoparticles in tissues from Sprague-Dawley rats. Part Fibre Toxicol 10(1):1–14. https://doi.org/10.1186/1743-8977-10-36

Lee JA, Kim MK, Song JH, Jo MR, Yu J, Kim KM, Kim YR, Oh JM, Choi SJ (2017) Biokinetics of food additive silica nanoparticles and their interactions with food components. Colloids Surf B: Biointerfaces 150:384–392. https://doi.org/10.1016/j.colsurfb.2016.11.001

Li Y, Cummins E (2020) Hazard characterization of silver nanoparticles for human exposure routes. J Environ Sci Health A Tox Hazard Subst Environ Eng 55(6):704–725. https://doi.org/10.1080/10934529.2020.1735852

Li J, Wuliji O, Li W, Jiang ZG, Ghanbari HA (2013) Oxidative stress and neurodegenerative disorders. Int J Mol Sci 14(12):24438–24475. https://doi.org/10.3390/ijms141224438

Li X, Liu B, Le Li X, Li YX, Sun MZ, Chen DY, Zhao X, Feng XZ (2014) SiO2 nanoparticles change colour preference and cause Parkinson's-like behaviour in zebrafish. Sci Rep 4:3810. https://doi.org/10.1038/srep03810

Li J, Yang S, Lei R, Gu W, Qin Y, Ma S, Chen K, Chang Y, Bai X, Xia S, Wu C, Xing G (2018) Oral administration of rutile and anatase TiO2 nanoparticles shifts mouse gut microbiota structure. Nanoscale 10(16):7736–7745. https://doi.org/10.1039/c8nr00386f

Li X, Ji X, Wang R, Zhao J, Dang J, Gao Y, Jin M (2020) Zebrafish behavioral phenomics employed for characterizing behavioral neurotoxicity caused by silica nanoparticles. Chemosphere 240:124937. https://doi.org/10.1016/j.chemosphere.2019.124937

Limón-Pacheco JH, Jiménez-Barrios N, Déciga-Alcaraz A, Martínez-Cuazitl A, Mata-Miranda MM, Vázquez-Zapién GJ, Pedraza-Chaverri J, Chirino YI, Orozco-Ibarra M (2020) Astrocytes are more vulnerable than neurons to silicon dioxide nanoparticle toxicity in vitro. Toxics 8(3): 51. https://doi.org/10.3390/TOXICS8030051

Liu F, Mahmood M, Xu Y, Watanabe F, Biris AS, Hansen DK, Inselman A, Casciano D, Patterson TA, Paule MG, Slikker W, Wang C (2015) Effects of silver nanoparticles on human and rat embryonic neural stem cells. Front Neurosci 9:1–9. https://doi.org/10.3389/fnins.2015.00115

Lomer MCE, Thompson RPH, Powell JJ (2002) Fine and ultrafine particles of the diet: influence on the mucosal immune response and association with Crohn's disease. Proc Nutr Soc 61(1): 123–130. https://doi.org/10.1079/pns2001134

Ma L, Liu J, Li N, Wang J, Duan Y, Yan J, Liu H, Wang H, Hong F (2010) Oxidative stress in the brain of mice caused by translocated nanoparticulate TiO2 delivered to the abdominal cavity. Biomaterials 31(1):99–105. https://doi.org/10.1016/j.biomaterials.2009.09.028

Ma W, Jing L, Valladares A, Mehta SL, Wang Z, Andy Li P, Bang JJ (2015) Silver nanoparticle exposure induced mitochondrial stress, caspase-3 activation and cell death: amelioration by sodium selenite. Int J Biol Sci 11(8):860–867. https://doi.org/10.7150/ijbs.12059

Mao BH, Chen ZY, Wang YJ, Yan SJ (2018) Silver nanoparticles have lethal and sublethal adverse effects on development and longevity by inducing ROS-mediated stress responses. Sci Rep 8(1):1–16. https://doi.org/10.1038/s41598-018-20728-z

Mata R, Nakkala JR, Chandra VK, Raja K, Sadras SR (2018) In vivo bio-distribution, clearance and toxicity assessment of biogenic silver and gold nanoparticles synthesized from Abutilon indicum in Wistar rats. J Trace Elem Med Biol 48:157–165. https://doi.org/10.1016/j.jtemb.2018.03.015

McCracken C, Zane A, Knight DA, Dutta PK, Waldman WJ (2013) Minimal intestinal epithelial cell toxicity in response to short- and long-term food-relevant inorganic nanoparticle exposure. Chem Res Toxicol 26(10):1514–1525. https://doi.org/10.1021/tx400231u

Mirsattari SM, Hammond RR, Sharpe MD, Leung FY, Young GB (2004) Myoclonic status epilepticus following repeated oral ingestion of colloidal silver. Neurology 62(8):1408–1410. https://doi.org/10.1212/01.WNL.0000120671.73335.EC

Mohammadipour A, Fazel A, Haghir H, Motejaded F, Rafatpanah H, Zabihi H, Hosseini M, Bideskan AE (2014) Maternal exposure to titanium dioxide nanoparticles during pregnancy; impaired memory and decreased hippocampal cell proliferation in rat offspring. Environ Toxicol Pharmacol 37(2):617–625. https://doi.org/10.1016/j.etap.2014.01.014

Murugadoss S, Lison D, Godderis L, Van Den Brule S, Mast J, Brassinne F, Sebaihi N, Hoet PH (2017) Toxicology of silica nanoparticles: an update. Arch Toxicol 91(9):2967–3010. https://doi.org/10.1007/s00204-017-1993-y

Musial J, Krakowiak R, Mlynarczyk DT, Goslinski T, Stanisz BJ (2020) Titanium dioxide nanoparticles in food and personal care products—what do we know about their safety? Nanomaterials 10(6):1–23. https://doi.org/10.3390/nano10061110

Neurotoxicity. Identifying and controlling poisons in the nervous system (1993) Environ Int 19(2): 213–214. https://doi.org/10.1016/0160-4120(93)90375-r

Pele LC, Thoree V, Bruggraber SFA, Koller D, Thompson RPH, Lomer MC, Powell JJ (2015) Pharmaceutical/food grade titanium dioxide particles are absorbed into the bloodstream of human volunteers. Part Fibre Toxicol 12:26. https://doi.org/10.1186/s12989-015-0101-9

Peters R, Kramer E, Oomen AG, Herrera Rivera ZE, Oegema G, Tromp PC, Fokkink R, Rietveld A, Marvin HJP, Weigel S, Peijnenburg AACM, Bouwmeester H (2012) Presence of nano-sized silica during in vitro digestion of foods containing silica as a food additive. ACS Nano 6(3): 2441–2451. https://doi.org/10.1021/nn204728k

Peters RJB, Undas AK, Memelink J, van Bemmel G, Munniks S, Bouwmeester H, Nobels P, Schuurmans W, van der Lee MK (2018) Development and validation of a method for the detection of titanium dioxide particles in human tissue with single particle ICP-MS. Curr Trends Anal Bioanal Chem 2(1):74–84. https://doi.org/10.36959/525/442

Peters RJB, Oomen AG, van Bemmel G, van Vliet L, Undas AK, Munniks S, Bleys RLAW, Tromp PC, Brand W, van der Lee M (2020) Silicon dioxide and titanium dioxide particles found in human tissues. Nanotoxicology 14(3):420–432. https://doi.org/10.1080/17435390.2020.1718232

Pinget G, Tan J, Janac B, Kaakoush NO, Angelatos AS, O'Sullivan J, Koay YC, Sierro F, Davis J, Divakarla SK, Khanal D, Moore RJ, Stanley D, Chrzanowski W, Macia L (2019) Impact of the food additive titanium dioxide (e171) on gut microbiota-host interaction. Front Nutrition 6:57. https://doi.org/10.3389/fnut.2019.00057

Rodríguez-Campuzano AG, Hernández-Kelly LC, Ortega A (2020) Acute exposure to SiO2 nanoparticles affects protein synthesis in Bergmann glia cells. Neurotox Res 37(2):366–379. https://doi.org/10.1007/s12640-019-00084-0

Rompelberg C, Heringa MB, van Donkersgoed G, Drijvers J, Roos A, Westenbrink S, Peters R, van Bemmel G, Brand W, Oomen AG (2016) Oral intake of added titanium dioxide and its nanofraction from food products, food supplements and toothpaste by the Dutch population. Nanotoxicology 10(10):1404–1414. https://doi.org/10.1080/17435390.2016.1222457

Shim KH, Jeong KH, Bae SO, Kang MO, Maeng EH, Choi CS, Kim YR, Hulme J, Lee EK, Kim MK, An SSA (2014) Assessment of ZnO and SiO2 nanoparticle permeability through and toxicity to the blood–brain barrier using Evans blue and tem. Int J Nanomedicine 9:225–233. https://doi.org/10.2147/IJN.S58205

Shimizu M, Tainaka H, Oba T, Mizuo K, Umezawa M, Takeda K (2009) Maternal exposure to nanoparticulate titanium dioxide during the prenatal period alters gene expression related to brain development in the mouse. Part Fibre Toxicol 6:2–9. https://doi.org/10.1186/1743-8977-6-20

So SJ, Jang IS, Han CS (2008) Effect of micro/nano silica particle feeding for mice. J Nanosci Nanotechnol 8(10):5367–5371. https://doi.org/10.1166/jnn.2008.1347

Song B, Liu J, Feng X, Wei L, Shao L (2015) A review on potential neurotoxicity of titanium dioxide nanoparticles. Nanoscale Res Lett 10(1). https://doi.org/10.1186/s11671-015-1042-9

Takahashi Y, Mizuo K, Shinkai Y, Oshio S, Takeda K (2010) Prenatal exposure to titanium dioxide nanoparticles increases dopamine levels in the prefrontal cortex and neostriatum of mice. J Toxicol Sci 35(5):749–756. https://doi.org/10.2131/jts.35.749

Takizawa Y, Hirasawa F, Noritomi E, Aida M, Tsunoda H, Uesugi S (1988) Oral ingestion of syloid to mice and rats and its chronic toxicity and carcinogenicity. Acta Med Biol 36(1):27–56

Tang J, Xiong L, Zhou G, Wang S, Wang J, Liu L, Li J, Yuan F, Lu S, Wan Z, Chou L, Xi T (2010) Silver nanoparticles crossing through and distribution in the blood-brain barrier in vitro. J Nanosci Nanotechnol 10(10):6313–6317. https://doi.org/10.1166/jnn.2010.2625

Temizel-Sekeryan S, Hicks AL (2020) Global environmental impacts of silver nanoparticle production methods supported by life cycle assessment. Resour Conserv Recycl 156:104676. https://doi.org/10.1016/j.resconrec.2019.104676

Trickler WJ, Lantz SM, Murdock RC, Schrand AM, Robinson BL, Newport GD, Schlager JJ, Oldenburg SJ, Paule MG, Slikker W, Hussain SM, Ali SF (2010) Silver nanoparticle induced blood-brain barrier inflammation and increased permeability in primary rat brain microvessel endothelial cells. Toxicol Sci 118(1):160–170. https://doi.org/10.1093/toxsci/kfq244

Umezawa M, Tainaka H, Kawashima N, Shimizu M, Takeda K (2012) Effect of fetal exposure to titanium dioxide nanoparticle on brain development—brain region information. J Toxicol Sci 37(6):1247–1252. https://doi.org/10.2131/jts.37.1247

Van der Zande M, Vandebriel RJ, Groot MJ, Kramer E, Herrera Rivera ZE, Rasmussen K, Ossenkoppele JS, Tromp P, Gremmer ER, Peters RJB, Hendriksen PJ, Marvin HJP, Hoogenboom RLAP, Peijnenburg AACM, Bouwmeester H (2014) Sub-chronic toxicity study in rats orally exposed to nanostructured silica. Part Fibre Toxicol 11(1):1–19. https://doi.org/10.1186/1743-8977-11-8

Van Kesteren PCE, Cubadda F, Bouwmeester H, Van Eijkeren JCH, Dekkers S, De Jong WH, Oomen AG (2015) Novel insights into the risk assessment of the nanomaterial synthetic amorphous silica, additive E551, in food. Nanotoxicology 9(4):442–452. https://doi.org/10.3109/17435390.2014.940408

Vance ME, Kuiken T, Vejerano EP, McGinnis SP, Hochella MF, Hull DR (2015) Nanotechnology in the real world: redeveloping the nanomaterial consumer products inventory. Beilstein J Nanotechnol 6(1):1769–1780. https://doi.org/10.3762/bjnano.6.181

Viau C, Haçariz O, Karimian F, Xia J (2020) Comprehensive phenotyping and transcriptome profiling to study nanotoxicity in C. elegans. PeerJ 8:e8684. https://doi.org/10.7717/PEERJ.8684

Vicente Lafuente J, Sharma A, Patnaik R, Muresanu DF, Shanker Sharma H (2012) Diabetes exacerbates nanoparticles induced brain pathology. CNS Neurol Disord Drug Targets 11(1):26–39. https://doi.org/10.2174/187152712799960808

Wang J, Zhou G, Chen C, Yu H, Wang T, Ma Y, Jia G, Gao Y, Li B, Sun J, Li Y, Jiao F, Zhao Y, Chai Z (2007) Acute toxicity and biodistribution of different sized titanium dioxide particles in mice after oral administration. Toxicol Lett 168(2):176–185. https://doi.org/10.1016/j.toxlet.2006.12.001

Wang J, Rahman MF, Duhart HM, Newport GD, Patterson TA, Murdock RC, Hussain SM, Schlager JJ, Ali SF (2009) Expression changes of dopaminergic system-related genes in PC12 cells induced by manganese silver or copper nanoparticles. Neurotoxicology 30(6):926–933. https://doi.org/10.1016/j.neuro.2009.09.005

Wang Y, Xiong L, Tang M (2017) Toxicity of inhaled particulate matter on the central nervous system: neuroinflammation, neuropsychological effects and neurodegenerative disease. J Appl Toxicol 37(6):644–667. https://doi.org/10.1002/jat.3451

Wei J, Liu J, Liang S, Sun M, Duan J (2020) Low-dose exposure of silica nanoparticles induces neurotoxicity via neuroactive ligand–receptor interaction signaling pathway in zebrafish embryos. Int J Nanomedicine 15:4407–4415. https://doi.org/10.2147/IJN.S254480

Węsierska M, Dziendzikowska K, Gromadzka-Ostrowska J, Dudek J, Polkowska-Motrenko H, Audinot JN, Gutleb AC, Lankoff A, Kruszewski M (2018) Silver ions are responsible for memory impairment induced by oral administration of silver nanoparticles. Toxicol Lett 290:133–144. https://doi.org/10.1016/j.toxlet.2018.03.019

Williams K, Milner J, Boudreau MD, Gokulan K, Cerniglia CE, Khare S (2015) Effects of subchronic exposure of silver nanoparticles on intestinal microbiota and gut-associated immune responses in the ileum of Sprague-Dawley rats. Nanotoxicology 9(3):279–289. https://doi.org/10.3109/17435390.2014.921346

Winkler HC, Suter M, Naegeli H (2016) Critical review of the safety assessment of nano-structured silica additives in food. J Nanobiotechnol 14(1):1–9. https://doi.org/10.1186/s12951-016-0189-6

Wu J, Wang C, Sun J, Xue Y (2011) Neurotoxicity of silica nanoparticles: brain localization and dopaminergic neurons damage pathways. ACS Nano 5(6):4476–4489. https://doi.org/10.1021/nn103530b

Xie H, Wu J (2016) Silica nanoparticles induce alpha-synuclein induction and aggregation in PC12-cells. Chem Biol Interact 258:197–204. https://doi.org/10.1016/j.cbi.2016.09.006

Xu F, Piett C, Farkas S, Qazzaz M, Syed NI (2013) Silver nanoparticles (AgNPs) cause degeneration of cytoskeleton and disrupt synaptic machinery of cultured cortical neurons. Mol Brain 6(1):29. https://doi.org/10.1186/1756-6606-6-29

Xu L, Shao A, Zhao Y, Wang Z, Zhang C, Sun Y, Deng J, Chou LL (2015) Neurotoxicity of silver nanoparticles in rat brain after intragastric exposure. J Nanosci Nanotechnol 15(6):4215–4223. https://doi.org/10.1166/jnn.2015.9612

Xue Y, Wu J, Sun J (2012) Four types of inorganic nanoparticles stimulate the inflammatory reaction in brain microglia and damage neurons in vitro. Toxicol Lett 214(2):91–98. https://doi.org/10.1016/j.toxlet.2012.08.009

Yamashita K, Yoshioka Y, Higashisaka K, Mimura K, Morishita Y, Nozaki M, Yoshida T, Ogura T, Nabeshi H, Nagano K, Abe Y, Kamada H, Monobe Y, Imazawa T, Aoshima H, Shishido K, Kawai Y, Mayumi T, Tsunoda SI et al (2011) Silica and titanium dioxide nanoparticles cause pregnancy complications in mice. Nat Nanotechnol 6(5):321–328. https://doi.org/10.1038/nnano.2011.41

Yoshida T, Yoshioka Y, Takahashi H, Misato K, Mori T, Hirai T, Nagano K, Abe Y, Mukai Y, Kamada H, Tsunoda S-I, Nabeshi H, Yoshikawa T, Higashisaka K, Tsutsumi Y (2014) Intestinal absorption and biological effects of orally administered amorphous silica particles. Nanoscale Res Lett 9(1):1–7. https://doi.org/10.1186/1556-276X-9-532

Younes M, Aggett P, Aguilar F, Crebelli R, Dusemund B, Filipič M, Frutos MJ, Galtier P, Gott D, Gundert-Remy U, Kuhnle GG, Leblanc JC, Lillegaard IT, Moldeus P, Mortensen A, Oskarsson A, Stankovic I, Waalkens-Berendsen I, Woutersen RA et al (2018) Re-evaluation of silicon dioxide (E 551) as a food additive. EFSA J 16(1):1–70. https://doi.org/10.2903/j.efsa.2018.5088

Younes M, Aquilina G, Castle L, Engel K, Fowler P, Frutos Fernandez MJ, Fürst P, Gundert-Remy U, Gürtler R, Husøy T, Manco M, Mennes W, Moldeus P, Passamonti S, Shah R, Waalkens-Berendsen I, Wölfle D, Corsini E et al (2021) Safety assessment of titanium dioxide (E171) as a food additive. EFSA J 19(5):e06585. https://doi.org/10.2903/j.efsa.2021.6585

Zane A, McCracken C, Knight DA, Young T, Lutton AD, Olesik JW, Waldman WJ, Dutta PK (2015) Uptake of bright fluorophore core-silica shell nanoparticles by biological systems. Int J Nanomedicine 10:1547–1567. https://doi.org/10.2147/IJN.S76208

Ziemińska E, Stafiej A, Struzyńska L (2014) The role of the glutamatergic NMDA receptor in nanosilver-evoked neurotoxicity in primary cultures of cerebellar granule cells. Toxicology 315(1):38–48. https://doi.org/10.1016/j.tox.2013.11.008

The manufacturer's authorised representative in the EU is Springer Nature Customer Service Centre GmbH, Europaplatz 3, 69115 Heidelberg, Germany. If you have any concerns regarding our products, please contact ProductSafety@springernature.com

Printed and bound by CPI Group (UK) Ltd, Croydon, CR0 4YY
01/10/2026
02226525-0001